God Does Play Dice With the Universe

A Startling New Picture of the World

Einstein Could Not Believe

But You Can Understand

Shan Gao

Published 2008 by arima publishing

www.arimapublishing.com

ISBN 978-1-84549-266-3

© Shan Gao 2008

All rights reserved

This book is copyright. Subject to statutory exception and to provisions of relevant collective licensing agreements, no part of this publication may be reproduced, stored in a retrieval system, or transmitted in any form or by any means, without the prior written permission of the author.

Printed and bound in the United Kingdom

Typeset in Times New Roman

This book is sold subject to the conditions that it shall not, by way of trade or otherwise, be lent, re-sold, hired out, or otherwise circulated without the publisher's prior consent in any form of binding or cover other than that which it is published and without a similar condition including this condition being imposed on the subsequent purchaser.

abramis is an imprint of arima publishing

arima publishing
ASK House, Northgate Avenue
Bury St Edmunds, Suffolk IP32 6BB
t: (+44) 01284 700321

www.arimapublishing.com

In memory of my ancestors,

who wrote the great Classic of Changes

I want to know God's thoughts; the rest are details.

—Albert Einstein

Contents

Illustrations	vii
Acknowledgements	viii
Preface	ix
The Riddle of Arrow	1
The flying arrow cannot be moving	1
Is motion an illusion?	2
The sands of time	5
The immobile picture of motion	9
Seeing is Believing?	11
Question of continuous motion	11
The mysterious double slit	15
What does modern science say?	18
The Cause of Motion	23
Force induces motion?	23
Motion has no cause	27
God Plays Dice	31
A logical road	31
Evidences	35
A mathematical viewpoint	41
Two seers	45
Time Division Universe	48
A particle is a lone cloud	49
The particle cloud ripples	56
Two fabrics	59
Psi wave and its evolution	64
The universe is an inseparable whole	68
The Gambling Rule	76
Where does the randomness go?	77
Playing dice in discrete space and time	80
God's gambling rule	86

The unification of two worlds	90
The Prime Mover	95
Causality and chance	95
Why does God play dice?	99
Who is God?	101
Notes	104
Bibliography	116
Index	119

Illustrations

Figure 1.1 The flying arrow cannot be moving 2
Figure 1.2 The trajectory of a moving ball 6
Figure 1.3 Atoms of time 8
Figure 1.4 The motion of a strange loop in a film 10
Figure 2.1 Is motion continuous? 12
Figure 2.2 A double-slit pattern 16
Figure 2.3 A mixture of two one-slit patterns 17
Figure 4.1 I don't know how to move 32
Figure 4.2 Continuous motion and discontinuous motion 42
Figure 5.1 A particle cloud 51
Figure 5.2 Electron clouds in a hydrogen atom 55
Figure 5.3 Two branches of a basic cloud superpose and ripple 57
Figure 5.4 Chinese Yin-Yang Diagram 63
Figure 5.5 The visual evolution of a particle cloud 67
Figure 5.6 The forming process of two entangled electrons 69
Figure 5.7 Two entangled electrons at six neighboring instants 70
Figure 5.8 Time division multiplexing 71
Figure 5.9 The picture of Schrödinger's cat 72
Figure 6.1 Randomness appears in discrete space and time 79
Figure 6.2 How far is it to the minimum size of space? 81

Acknowledgements

The ideas of this book come out of my lonely exploration in the past twenty years. I have also benefited from discussions with many physicists who care about the way the world really is. They are: Samuel Braunstein, ZeXian Cao, Bernard d'Espagnat, Alwyn van der Merwe, Philip Pearle, Roger Penrose, Abner Shimony, Antoine Suarez et al. I thank them all deeply. At the same time, I am very grateful to my parents, QingFeng Gao and LiHua Zhao, who brought me up and gave me a good education, and my wife, HuiXia Liu. This book could not have been completed without their care and support. Finally, I am grateful to my lovely daughter RuiQi. She is full of curiosity and always asks me some naïve questions about the universe. They let me constantly rethink the accepted picture of the world. She is very like I was during childhood!

Preface

We live in a classical world. Yet, there is a ghostlike atomic world underneath. Everyone knows that a ball is composed of atoms. But nobody knows what atoms look like, and especially, how in hell atoms move. They don't look like tiny balls at all; for instance, a single atom can pass through two slits at the same time. The more stunning fact is that we don't exactly know how a ball moves either. It appears to move in a continuous way. This, however, is a mere illusion. Even the greatest scientists Newton and Einstein were also deluded by the appearance. How on earth do objects move then?

This book will reveal a deep secret of nature for the first time. It is that everything in the universe, whether it is an atom or a ball or even a star, ceaselessly jumps in a random and discontinuous way. In a famous metaphor, God does play dice with the universe. This picture of reality is so strange that nobody even dreamed of it. But it is real. Discovering that motion is not continuous but discontinuous and random is like finding the Earth is not at rest but moving. It will lead to a profound shift in our world view. Now we can finally walk out of Plato's cave, and approach the light in the real world. Reality is really amazing!

During my childhood, it had been a wonder for me that the twinkling stars strewed in the night sky don't fall to the Earth. I had a strong desire to know the whys and wherefores. Later I found the answer in textbooks. It changed my picture of the universe. When I was an undergraduate, I was entranced by the deep mysteries of the atomic world. I was especially stunned by the fact that the commonsensible planetary picture of atoms turns out to be utterly false; the electron in an atom cannot rotate round the atomic nucleus as the Earth rotates round the sun, or else it would soon radiate its energy and fall into the nucleus, and as a result, my body composed of atoms would collapse in a blink. How does the electron move then? It must exist in the atom. It must move in some way there. But more surprisingly, textbooks provided no picture of the motion of an electron. On 22 August 1987, I wrote in my diary: "Is it really true that we have no way to describe the atomic processes as processes happening in space and time?" I could but search for the answer by myself. Then I started on a lonely journey to "trace" the elusive electron at the age of 16.

In order to find how the electron moves in an atom, I went to the Institute of Electronics, Chinese Academy of Sciences to pursue my graduate study. But it was according to expectation that nobody there could give me any tips either. I then spent nearly every day in musing the seemingly indescribable motion picture of electrons. If a ball indeed moves in a continuous way, then it seems that an electron or an atom should also move in the same way. The ball is composed of atoms after all. But, on the other hand, if an electron moves continuously in an atom, it will soon fall into the nucleus, while the tragedy does not happen in reality. This is a great dilemma. I found some possible solutions, but they shortly proved to be wrong.

The puzzle had been plaguing me. Day after day, I gradually doubted the reality of continuous motion. But I still felt in my bones that the particle must move in some way. Finally, in the early morning of 12 October 1993, I experienced a sudden enlightenment. At that moment, I felt that my body permeated the whole space of the universe and I was united with it. "I" disappears. A clear picture then appeared: a particle is jumping in a random and discontinuous way. It is not inert but active; it moves purely by its own "free will". God told me He plays dice in the atomic world. I finally broke loose the tightest shackles of continuous motion with the help of inspiration. I could then see the true face of motion. After the event, this outcome seems very natural from a logical point of view. Since a particle cannot move continuously, it must move in a discontinuous way. How deep-rooted the prejudice of the uniqueness of continuous motion is!

If an atom moves in a random and discontinuous way, then it can easily pass through two slits at the same time. But why does a ball appear to move in a contrary way? Moreover, why in hell does God play dice? These puzzles further haunted me. Yet, no institute or college would support a researcher interested in these seemingly fantastic problems. So I decided to be an independent research scientist, or more accurately, a natural philosopher who aims at understanding the mysterious universe. Life was not easy. But I never gave up my research, and I never stop thinking. It had become the theme of my life. Curiosity then turned to responsibility. I must understand God's thoughts. I must let all people see the light of truth.

As time went on, the picture of random discontinuous motion became clearer and clearer in my mind. When I took a walk one afternoon in June 2001, I suddenly had another inspiration after long reflection in solitude and meditation. I realized that motion has no cause in reality, and thus it must essentially be random, i.e., God must play at dice. Moreover, the familiar inertial motion of a ball has actually revealed that it also jumps in a random and discontinuous way just like an atom. This is a great revelation. Maybe the path to truth is always devious in order that surprise can hide at the turn waiting for persevering seekers. God also plays dice in our classical world. He actually plays dice with the whole universe. What a harmonic picture of the world!

I simply want to know the answer of a naïve question. I simply think on it continually. But the exploration has completely changed my life. It shapes the way of my life, and finally leads me to God, the ultimate reality. As Trinity said in *The Matrix*, "It's the question that brought you here… The answer is out there, Neo, and it's looking for you, and it will find you if you want it to."

Truth is simple. But in order to explain it in plain language, it becomes a little complex so that a mini-book is needed. No equations. No jargon. There is only a clear and amazing picture of the universe here. It is comprehensible to everyone. Especially, no knowledge of quantum physics is needed. In fact, the book will lead you from our familiar classical world to the weird atomic world along a logical road. You can then understand the enigmatic quantum more deeply than its discoverers. The ultimate truth will be simple and apprehensible.

Life is transitory. Everybody is a mere mote in the universe. Yet God gives us mind; thus we can know and understand His thoughts. The most happiness is not beyond this. As the great Chinese sage Confucius taught us in *The Analects*, "Hear the Tao in the morning, and it would be all right to die that evening." I hope that this book will not only tell you a startling new picture of the world, but also make your life more colorful.

Shan Gao
Beijing
October 2007

When he approaches the light his eyes will be dazzled, and he will not be able to see anything at all of what are now called realities... He will require to grow accustomed to the sight of the upper world. And first he will see the shadows best, next the reflections of men and other objects in the water, and then the objects themselves; then he will gaze upon the light of the moon and the stars and the spangled heaven... Last of he will be able to see the sun.

—Plato, *The Republic*

1
The Riddle of Arrow

Time flies like an arrow.
—An old adage

Motion is *the* eternal subject of human enquiry. The study of motion begins with an old and famous arrow. Its owner was the Greek philosopher Zeno of Elea, who lived about 2500 years ago.

The flying arrow cannot be moving

Zeno was the first man who seriously pondered over the puzzle of motion. He conceived many paradoxes of motion, the most famous being that of the arrow.[1] To our great surprise, he argued that motion is actually an illusion. But was he right?

Imagine an arrow in flight. At every instant in time, it is located at a specific position. If the instant is durationless, then the arrow does not have time to move and is at rest during that instant. Now, during the subsequent instants, it must also be motionless for the same reason. Since the entire period of motion consists only of instants, all of which contain the arrow at rest, Zeno concludes, the flying arrow is always at rest and cannot be moving: its motion is merely an illusion.

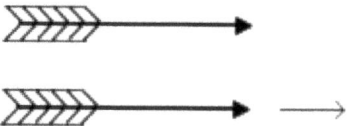

Figure 1.1 The flying arrow cannot be moving

This conclusion still holds true when the instant has finite duration, which is the smallest part of time. Suppose that the arrow actually moved during such an instant. It would be at different locations at the start and end of the instant, which implies that the instant has a 'start' and an 'end', which in turn implies that it contains at least two parts, and thus is not the smallest part of time at all. This leads to a contradiction. So, the flying arrow cannot be moving even when the instant has finite duration.

"No, no, no, the flying arrow must be moving!" you probably cannot help crying out. But what's wrong with Zeno's argument? Don't be afraid of philosophers. At any rate, there are only two possible answers: one is that Zeno was right, and motion is indeed an illusion; the other is that Zeno's argument is wrong. Let us first look at the former.

Is motion an illusion?

> *Have you ever had a dream, Neo, that you were so sure was real? What if you were unable to wake from that dream? How would you know the difference between the dream world and the real world?*
> —Morpheus in *The Matrix*

We had better doubt our perceptions before we believe them, as we are dreamful beings. With respect to our perceptions of motion, we also need to reflect on them before accepting the reality of motion.

If Zeno is right, then everything existing in time would be still. The sun would not rise and set and the moon would not wane and wax either. Your eyes would not blink, and your heart would not beat either. Moreover, your perceptions and your mind would all be at rest. No doubt, this would be the greatest revelation if it were true. As in the film *The Matrix*, we are all deceived by our perceptions. Those who lived in the Matrix world thought they had free will, yet they were actually designed and controlled by the Matrix program. It might be possible that our perceptions also deceive us concerning motion; they tell us that motion is real, but motion is actually an illusion. Is this really true?

It seems motion can be generated by both our body and our mind. As we know, motion is relative. Let an object be initially at rest relative to you. When you begin to move, you will observe that the object also begins to move. Yet no external force results in the motion of the object. Its motion is generated by your motion in a sense. In addition, our mind can also generate the illusion of motion. We sometimes dream of flying in the air. But we do not actually fly; the flying is only an illusion generated by our mind. So, why is it impossible that all motion is just an illusion?

Suppose everything including our mind is motionless. How can the illusion of motion be generated by our mind then? It seems that two immobile things cannot generate a moving thing in nature. This is very similar to the situation where the sum of two zeros is still zero. As a consequence, even though the motion of external objects is an illusion, our mind itself must move. Modern science reveals that our brain contains a large number of atoms in constant motion. In order to generate the illusion of motion, these atoms must undergo real motion.

But what is mind? Is it composed of material atoms? These are unsolved puzzles in reality. As René Descartes famously said, "I think, therefore I am." [2] We do know the existence of mind. Yet we haven't exactly known its relation with matter. What we can access and trust is only our mind, and everything in our world is its construction. As a result, our belief in the reality of an external world cannot be justified. We can but postulate its existence. This is the well-known doctrine of David Hume, a British skeptic.[3] Certainly, the external world may probably exist. But its existence is simply beyond our ability to prove. Aha, how powerful and powerless our mind is!

Don't be too heart-struck. At all events, motion does exist. Even if the whole external world doesn't move or even doesn't exist, our mind still exists and does move in its space. We still need to live on in such a strange world. Especially, apparent motion also exists, and it has effects. For instance, if someone falls off a skyscraper, his mind will cease moving forever. Therefore, it is still meaningful to study the apparent motion. Furthermore, if we understand it, we will also lead a better life. In fact, as most people believe, a real external world in constant motion exists in all probability. In Confucius's famous aphorism, "Everything flows on and on like the river, without pause, day and night." [4]

Now that motion is real, Zeno's argument must be wrong. Where is the mistake then?

The sands of time

The sands are number'd that make up my life.
—William Shakespeare, *King Henry VI*

As we can see, Zeno's argument strongly relies on the premise that there exist individual instants during which the arrow is motionless. This premise seems very natural. The picture of time as a continuum of durationless instants also accords with our common sense of time. But is it true? Do instants really exist?

In everyday life, we know the instant notion very well. Every clock has a numbered dial indicating instants. Our language is also full of instantaneous descriptions such as getting up at 8 o'clock in the morning. Moreover, the common use of the instant notion is supported by science, especially by the existence of points in math, the most precise science. As we have learned from elementary math, a line is composed of points, each of which can be represented by a real number. In fact, the point concept is a foundation stone of math. In the set theory, which is the basis of modern math, everything is a set composed of individual elements. As a typical example, a real line is a nondenumerable point set. In this way, a line can be built up from dimensionless points.

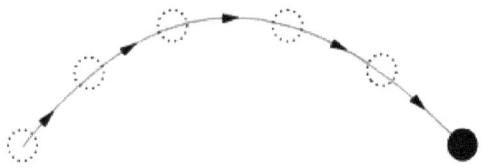

Figure 1.2 The trajectory of a moving ball

Since math is used to describe the physical world, the instant concept also plays a basic role in physics. It is a standard assumption that time is composed of durationless instants, and a moving body is in one position in space at each instant during the course of motion. For instance, a moving ball is in one position in space at each instant during its motion. These positions constitute its trajectory, which is described by a line consisting of points in math. Besides, it is a well-known fact that the positions of the moving second hand of a clock just represent instants.

However, although the instant notion is indispensable in modern science, the durationless instants are inaccessible by experience. Only finite time intervals can be measured. If something cannot be detected, or

---

The shortest time interval ever measured

A group led by Ferenc Krausz of Vienna University of Technology used pulses of laser light to watch electrons moving around atoms, and were able to distinguish events that took place 100 attoseconds – or 10^{-16} seconds – apart.

—Mark Peplow, *Nature,* February 26, 2004

---

in more objective language, cannot display itself in essence, then it seems reasonable that it cannot exist. Such a view is called minimum ontology in philosophy. Most scientists are not disturbed by this worry. But a few of them have indeed questioned the reality of durationless instants. They prefer the so-called gunk view, according to which time and space don't consist of durationless instants and positions; rather, they are infinitely divisible like gunk.

The gunk concept can be traced back to Aristotle. He used it to refute Zeno's paradox of the arrow. If durationless instants don't exist at all, then Zeno's argument will naturally collapse. In fact, motion is simply unanalyzable in the gunk world. However, nobody has successfully constructed a new math based on the gunk concept as yet. It seems that such math must also be constructed on the basis of standard math. In addition, the gunk lovers also encounter serious problems in the physical world. It is very difficult to describe continuous variations in terms of the gunk concept. As Isaac Barrow, the mentor of Newton, noted in his *Mathematical Lectures* (1734), "Rest is often peculiar to them (i.e. points)…as…to the center of a wheel."[5] In the gunk world, however, there is no such center. So every part of the wheel must move! This seems very absurd. Certainly, we can invent a more complex theory to account for such simple phenomena. But, as some contemporary gunk lovers admit, the gunk view seems implausible when considering the simplicity and naturalness of physical theory.[6]

If we insist that unobservable things cannot exist, then the gunk view should also be rejected. The assumed infinite divisibility of space-time cannot be confirmed either; we can never detect an infinitesimal space-time region. What is left for our notion of time then? Time is composed of instants with finite duration. Such instants are the smallest parts of time, which may be called temporal atoms. Since a finite time interval can have effects and be measured in principle, this notion of time can be examined by experience. We will discuss it in detail later on.

Figure 1.3 Atoms of time

Now let us come back to the arrow paradox. If instants really exist, Zeno's premise will be valid. But we also know motion is indeed real. Then what mistake did Zeno make? How can motion emerge out of motionless instants alone? As we will see, although the picture of motion is much stranger than we usually think, it is very familiar to the moviemakers.

The immobile picture of motion

> *A sequence of photographs projected onto a screen with sufficient rapidity as to create the illusion of motion and continuity.*
> —'movie' in the *American Heritage Dictionary*

The standard answer is as follows. It is fallacious to conclude from the fact that the arrow doesn't travel any distance in an instant that it is at rest. Motion has nothing at all to do with what happens during instants; it has instead to do with what happens between instants. In short, motion is merely a feature of being in different locations at different times, and that is that. If an object has the same location at the instants immediately neighboring, then we say it is at rest; otherwise it is in motion. This view is also called the 'at-at' theory of motion. Therefore, since the arrow in flight has different positions at different instants, it is surely moving.

We usually think that a moving object should move at every instant. The actual picture of motion, however, is that a moving object does not move during any instant. In Henry Bergson's memorable words, "movement is composed of immobilities." [7] This consequence is very counterintuitive. Yet it discloses the first secret of motion. Then when does the flying arrow actually move? How does it get from one place to another? There is only one answer: the arrow moves from one position to another simply in virtue of being in different positions at different times. As Bertrand Russell clearly put it: "Motion consists merely in the occupation of different places at different times." [8]

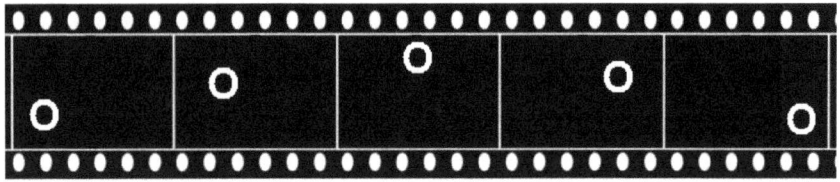

Figure 1.4 The motion of a strange loop in a film

It is a solid experiential fact that motion does exist. Zeno's arrow further teaches us that motion is merely being in different positions at different times. So the motion of an object is really like a movie; the object being in one position is just one frame of the movie. But is the transition from one position to another indeed continuous as in a movie? In short, is motion continuous?

2
Seeing is Believing?

Nature makes no leaps.
—Gottfried Leibniz

If you believe motion is continuous, you are not alone. Both Newton and Einstein also thought so. But is motion really continuous? Does Nature indeed make no leaps?

Question of continuous motion

The true face of mountain Lu remains unknown to me,
It is simply because I myself am on the mountain.
—Written on the wall at West Forest Temple by Su Tungpo [1]

How on earth do objects move? This is a great puzzle. Most people may think motion is obviously continuous; this accords with our everyday experience of the motion of objects. In fact, when people talk about motion, they only refer to continuous motion. The words "path" and "trajectory" and so on in dictionary all imply the picture of continuous motion. But is continuous motion *the* real motion?

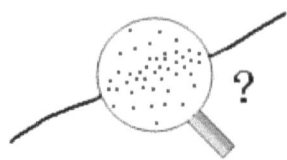

Figure 2.1 Is motion continuous?

We live in a classical world. We are only familiar with continuous motion after all, and thus we cherish it so deeply. We have been taking for granted that continuous motion is the only possible form of motion, as well as the real form of motion. At first glance, the existence of continuous motion seems very natural. An object will sustain its velocity if no influence is imposed on it, as there is no cause to change its velocity. Then the free object can but be at rest or move continuously with a constant velocity. Besides, a moving object is in one position at an instant, and it can only be in the neighboring positions at the adjacent instants. Again: no cause results in the jump of an object from one position to another non-adjacent position. Indeed, we never see a car jumping from one place to another without passing through in-between places to avoid a traffic jam. In a word, the existence of continuous motion seems inevitable. If it is not the real motion, then which form of motion is the real motion? Since we never see, never learn of and even never dream of another form of motion, how could it be the real motion?

> **Plato's cave**
>
> The Greek philosopher Plato tells the famous *Allegory of the Cave* in Book VII of *The Republic*. Suppose there are some men chained up to a wall in a cave. Behind them burns a fire. These men can see nothing but the shadows of objects behind them such as puppets shown by some puppeteers. Such prisoners would regard the shadows as real and would know nothing of the real causes of the shadows. Plato believes that we are like those men sitting in the cave: the world revealed by our senses is not the real world, which can only be apprehended intellectually.

Nature always hides her secret, however. Continuous motion is probably an illusion. Maybe the classical world we live in is just Plato's cave, and continuous motion is merely the shadow of real motion. When seeing a film, we also think that the objects in the movie move in a completely continuous way. Yet every movie is actually composed of discrete photographs, which are projected onto a screen with sufficient rapidity (e.g. at 24 frames per second). Because human cannot identify such rapid changes, movie can successfully create the illusion of continuous motion for filmgoers. Similarly, our everyday experience of continuous motion may also deceive us. So let us examine it more thoroughly.

Consider an object moves from point 0 to point 1 in a straight line. If the motion of the object is continuous, then it must pass through all points between 0 and 1 one by one. However, there are infinitely many points between 0 and 1, say 1/2, 1/4, 1/8 etc. We cannot count up to them during a finite time interval, and we cannot trace every position of the object either. Then how can we know the object really passes through all these points in a continuous way? If we cannot know, how can we confirm that the motion of the object is continuous?

There might exist some other methods to confirm the existence of continuous motion. For example, although we cannot directly verify that the object passes through all points between 0 and 1, we may probably demonstrate this conclusion through a plausible hypothesis. One possible hypothesis is that an object moving from one position to another must pass through their middle position. However, even though such a hypothesis can help to confirm the existence of continuous motion, how can we verify this hypothesis? It might be right for a large distance, but has it been verified for a very small distance? Since there also exists infinitely many different distances, the hypothesis cannot be confirmed by experiment either. In fact, even if the above hypothesis is verified, it cannot confirm the continuity of motion. For instance, the discontinuous trajectory consisting only of rational points evidently satisfies the above hypothesis. Thus, it seems that we cannot confirm the existence of continuous motion with the help of another hypothesis either.

In a word, it is only an unverifiable assumption or belief that motion is continuous. But this fact alone cannot refute the actual existence of continuous motion. Maybe reality is inaccessible in nature. In the final analysis, infinity gets in our way. In order to discover the real motion, we must enter into a smaller and smaller space, even an infinitesimal space. Then is there any evidence of the motion different from continuous motion?

The mysterious double slit

> *A phenomenon which is impossible, absolutely impossible,*
> *to explain in any classical way.*
> —Richard Feynman

Motion is probably much stranger than we usually think. We can only see the apparent motion of macroscopic objects such as a ball after all. How about the motion of microscopic particles such as an atom? Here we will discuss a famous experiment of particles, the double-slit experiment. The experiment is so simple that everyone can understand it. As we will see, however, it cannot be explained in terms of the picture of continuous motion.

In a typical double-slit experiment, the single particle (e.g. electron) is emitted from the source one after the other, and then passes through the two slits to arrive at the detecting screen. In this way, when a large number of particles with the same energy arrive at the screen, they collectively form an undulant double-slit pattern. The ridges in the pattern are formed in the positions where more particles reach, and the valleys in the pattern are formed in the positions where nearly no particles reach.

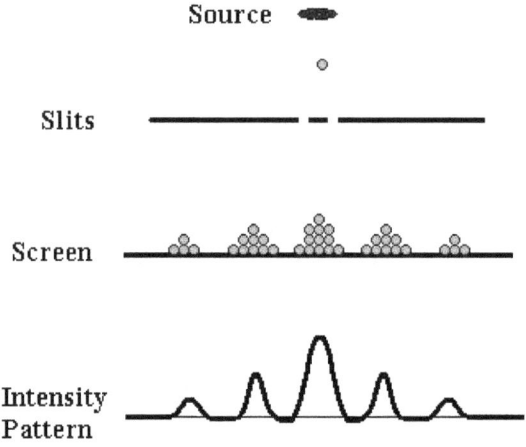

Figure 2.2 A double-slit pattern

According to the picture of continuous motion, the single particle can but pass through one of the two slits. One expects the double-slit pattern should be the same as the direct mixture of two one-slit patterns, each of which is formed by opening each of the two slits independently. The reason is that the passing process of each particle in a double-slit experiment is exactly the same as that in one of the one-slit experiments. For instance, if a particle passes through the left slit and then arrives at one position of the screen in the double-slit experiment, then this process will also happen in the same way in the one-slit experiment in which the right slit is shut.

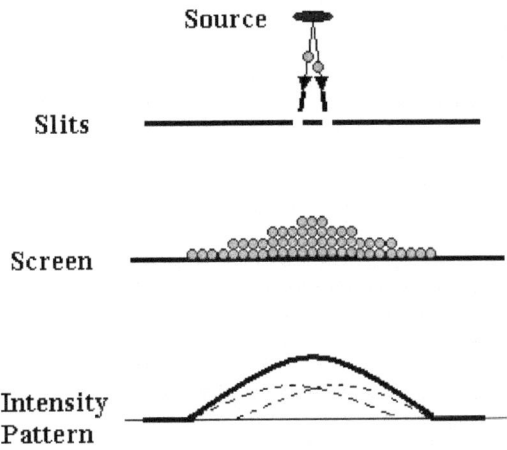

Figure 2.3 A mixture of two one-slit patterns

To our great surprise, however, experiments show that the patterns for the above two situations, namely the double-slit pattern and the mixture of two one-slit patterns, are obviously different. The difference cannot be explained in terms of the picture of continuous motion. In fact, we can see where the perplexity lies more obviously from the following fact, i.e., that when one of the two slits is shut, the particle can reach some positions on the screen (e.g. the positions where the valleys in the double-slit pattern are formed), but when the shut slit is opened, it will prevent the particle from reaching these positions on the screen. So it seems that a single particle must pass through both slits in the double-slit experiment, and its motion cannot be continuous.

Up to now, the double-slit experiment has been accomplished with different microscopic particles such as photons, electrons, neutrons, atoms, and even molecules. Therefore, it seems that the motion of small objects, which cannot be directly observed by our naked eyes, is not continuous but discontinuous. If you would like to take an objective attitude, you might also think so. The double-slit experiment seems to provide definitive evidence after all. Then what are the opinions of modern scientists?

What does modern science say?

Shut up and calculate.
—A slogan of modern scientists

How does a single particle pass through two slits in the double-slit experiment? It seems evident that experiments have revealed that the single particle passes through both slits in a discontinuous way. Therefore, the motion of particles is not continuous but discontinuous. To our surprise, however, modern science gives no definite answer to this simple question. More surprisingly, there are many different opinions among scientists, but none of them is the same as the above.

The main aim of modern science is no longer to understand the world. It is indeed very successful. For instance, it can send human to the moon, and it can also build a powerful atomic bomb. However, it cannot answer such a simple question, i.e., that how a single particle passes through two slits. Moreover, when you curiously ask a similar question, the standard response from a standard scientist is actually "shut up and calculate". As the Nobel Physics Laureate and acclaimed educator Richard Feynman once said seriously to his audience, "Do not keep saying to yourself, if you can possibly avoid it, 'But how can it be like that?' because you will get 'down the drain' into a blind alley from which nobody has yet escaped. Nobody knows how it can be like that."[2] Thus, it might be not unexpected that modern science cannot answer the above simple question. Now let us see what in hell it says.

The orthodox answer to the above question is that the question itself is meaningless. So you should not ask such a silly question. This point of view was mainly proposed by the Danish physicist Niels Bohr, and has been widely accepted by most scientists today. It must be very shocking when you hear this unbelievable answer for the first time. Indeed, Bohr had also warned us: "Those who are not shocked… cannot possibly have understood it."[3] You might immediately wonder why nearly all scientists accept this answer. So let us look at the double-slit experiment more closely.

According to the orthodox view, if you want to know how the single particle passes through two slits to form the double-slit pattern, you must detect which slit the particle passes through by taking a position measurement. Yet this kind of measurement will inevitably destroy the double-slit pattern. Then on condition that the double-slit pattern is not influenced, we cannot detect which slit the single particle passes through, and thus we cannot know how the single particle passes through the two slits in the double-slit experiment. Furthermore, the orthodox view insists that the realistic picture of the particle passing through two slits doesn't exist in essence, as we cannot detect it. Then the answer of the above question not only can never be known, but also doesn't exist at all. So, the question itself is absolutely meaningless.

> **Seeing destroys the seen**
>
> What does the cat hiding in the dark look like? The question seems meaningless. You cannot know what the cat in the dark looks like by looking at it. When you turn on the light, the cat is no longer that in the dark, but a *new* cat in the light. In an extreme situation, the light could kill the cat.

It is not easy to find the loopholes of the orthodox view.[4] Even its strongest opponent Einstein also admitted its logical consistency. He, however, utterly rejected it based on his scientific belief.[5] For him, there must exist an objective picture of motion for microscopic particles as well as for macroscopic objects. Moreover, Einstein strongly believed that motion is continuous, as Newton had told us. But a problem immediately appears: how can this unorthodox view explain the double-slit experiment? As we have seen, the picture of continuous motion cannot explain the experiment at all.

Unexpectedly, Einstein's follower David Bohm, an American

physicist, indeed found a way out.[6] Bohm assumed that microscopic particles still move continuously. This means that the single particle passes through only one of the two slits in the double-slit experiment. But in order to account for the formed double-slit pattern, he further assumed that the particle is always accompanied by an unusual wave. This wave has no energy, but it can guide the motion of the particle as a radar wave guides the motion of an active missile. In the double-slit experiment, the guiding wave passes through both slits and leads the particle to move along a proper trajectory to form the double-slit pattern. Such trajectories are continuous but not straight. At first glance, all these are very appealing. However, both the guiding wave and the continuous trajectory are unobservable in essence. Then in what sense they exist? This is a fatal flaw of Bohm's theory. In fact, even Einstein also thought it was too cheap.

There are also many other explanations of the double-slit experiment. Stranger and stranger words and pictures, say many worlds and many minds and so on, have been continually invented. It seems reasonable that there exist different explanations for a phenomenon in social science. Yet natural science should not be like that. Admittedly, the fact that scientists take many different positions – and students don't know which position to accept – marks an extremely embarrassing period in the history of science. Unfortunately, this is just the present situation of modern science.

You must be very disappointed at present science, as it cannot even answer very simple questions. How does a single particle pass through two slits? Is motion really continuous? Indeed, the explanation of the double-slit experiment is still in hot debate today. But one thing is certain; there must exist some deep-rooted prejudices in our present understanding of motion. It is just them that prevent us from discovering the real picture of motion. In order to move on before experience can help us, we have to come back to the origin of motion.

3

The Cause of Motion

Happy is he who can recognize the causes of things.
—Virgil, *Georgics*

Motion involves change in position. If we can find the actual cause of the change, we may discover how on earth objects move. This is a rather simple idea. To our great surprise, however, it is actually the key to unravel the great puzzle of motion.

Force induces motion?

Every event has a cause.
—An axiom

Aristotle was the first to seek the origin of motion fundamentally. He held that external force is the cause of motion, and there is no motion without a force. Especially, motion requires a force to sustain it; moving objects only continue to move so long as there is an external force inducing them to do so. This is intelligible, as it accords with the common sense of causality, i.e., that there is no effect without a cause.

However, Aristotle's theory is inconsistent with experience. For instance, an arrow keeps flying after the bowstring is no longer pushing on it. In order to account for such phenomena of continued motion, Aristotle assumed that the action of the surrounding air continues to move the projectile in some way. Unfortunately, this *ad hoc* explanation also contradicts experience. As an obvious example, when you sit on a sled and are pushed across ice, you will continue to move after the pushing is stopped, while you feel no pushing force from air during the continued motion.

In order to avoid the difficulty for explaining continued motion, the idea of impetus was presented notably by the medieval scholar Jean Buridan. He argued that a projectile continues in motion not, as Aristotle held, because it is pushed by the surrounding air, but because of the force transmitted to it by the agent that launched it. This force internal to objects is called impetus. So, according to Buridan, impetus is the cause of motion. The motion that occurs without an external force is sustained by an internal motive force, the impetus, which can be transferred from an external propelling agent that initializes the motion.

The impetus belief was very popular in the pre-Newton times. In fact, Newton was also its adherent at one time. He called it "the force of a body". More surprisingly, many contemporary students still have such a conviction.[1] The remarkable popularity of the impetus belief strongly implies that it has some reasonable elements. To begin with, this belief might be derived from a lifetime of kinesthetic experience, and seems

> **Contemporary misconceptions about force**
>
> A research in 1982, which was conducted by the American physicist J. Clement, showed that nearly 80% of a group of engineering freshmen had the impetus belief. They thought the force from hand pushes up on the tossed coin when it is on the way up.

consistent with everyday experience. Secondly, the belief is natural and intelligible. Motion involves change of position. A change requires a cause according to the common sense of causality. So motion must have a cause. Since continued motion (e.g. the flight of an arrow) occurs without an external force in the direction of motion, there must exist an internal motive force that moves the object in this direction. It is just the impetus. Because the change of position is primary for motion, this line of reasoning seems justifiable.

The impetus theory, however, is not consistent with experience either. The above example of sledging has provided a refutation; you feel no force during the continued motion of your sled. In addition, the theory cannot account for the relativity of motion either. It is actually a general consequence that a theory asserting motion has cause such as impetus contradicts the relativity of motion or the equivalence between motion and rest. Let us take an illustration. An object is initially at rest relative to you, and no force causes its motion. When you move with a constant velocity relative to the object, you will observe that the object also moves relative to you. Since no influence (e.g. impetus) is added to act upon the object during this process, the motion of the object must have no cause, and particularly impetus is not the cause of its motion.

The consideration of the equivalence between motion and rest also made Newton finally transform impetus into inertia, and further ignited the Newtonian revolution. That was a great conceptual change.[2] Let us see how Newton achieved the transformation.

In a short paper sent to Edmond Halley in the spring of 1685, Newton said: "The inherent and innate force of a body is the power by which it preserves in its state of rest or of moving uniformly in a straight line."[3] In a subsequent short paper sent to Halley he changed the expression 'The inherent and innate force of a body' to 'The internal force of matter', and explained that it is: "the power of resistance by means of which any one body continues so far as it can in its state of rest or moving uniformly in a straight line… not differs at all from the inertia of matter except in our mode of conceiving it."[4] Here Newton had transformed impetus into the modern concept of inertia. Then he further explained force as "an action exercised on a body to change its state of rest or motion. This force consists truly in the action only, nor does it remain in the body after the action."[5] As such, Newton founded classical mechanics, which gained huge success in the later 200 years.

In Newton's world, motion and rest are equivalent. An object can sustain its motion just as it can sustain its rest. No force is needed to sustain the motion of an object, and the object in a state of motion possesses an "inertia" that causes it to remain in that state of motion. This postulate is well summarized in Newton's first law of motion or the law of inertia.

The law of inertia perfectly accords with macroscopic experience, and it also appears logical and even self-evident. A freely moving object should hold its velocity, as there is no cause resulting in the change of its velocity. Thus the object must continuously move in a straight line with a constant speed, as the law of inertia requires. But has the law of inertia disclosed all secrets about the origin of motion?

Motion has no cause

Stirring of the dry bones.
—An idiom from the Bible

According to Newton's view, force is not the cause of motion but the cause of the change of motion. Motion itself needs no causal explanation; a freely moving object continues to move because it has inertia and thus can sustain its motion. Yet there is a big trap here. Nobody has come out from it as yet.

Let us think back Zeno's arrow. It has taught us that motion is essentially being in different positions at different times, and there is no motion at each instant. So no motion is available for an object to sustain at every instant in reality; motion and its velocity simply don't exist at instants.[6] Therefore, Newton's view is not wholly valid.

In a sense, the concept of inertia diverts our attention from position change, which is *the* essential characteristic of motion, to velocity change, which is merely an apparent characteristic of motion.[7] As a consequence, it evades the question about the cause of motion, and veils the true face of motion. This fact was neglected by Newton and all great men after him. The orthodox slogan is "we need only explain changes in motion, not motion itself." Yet it is actually a deep-rooted prejudice. Only after we reject it, can we open the door to the real world.

> **The orthodox answer**
> Q: What is the cause of motion?
> A: We need only explain changes in motion, not motion itself.

Although Newton's explanation of inertial motion is not right, his discovery can lead us to the correct one. According to Newton, neither external force nor internal force is the cause of motion. So there is only one left possibility, i.e., that motion has no cause. No cause determines how the position of an object changes in reality.[8] This is a great revelation. It teaches us again that asking a proper question is more important than answering an improper question. What is the cause of motion? It seems that nobody has seriously asked this question since Newton's times. But it is just this seemingly naïve question that leads us to the real origin of motion. Against all expectations, the answer is rather simple; motion has

3 – THE CAUSE OF MOTION

no cause.

Now we discover that objects are not inertial but active in essence. Inertial motion is only an apparent display of objects. In fact, an object can spontaneously change its position without a cause. This is a far-reaching conceptual change after the Newtonian revolution. As we will see, it will deeply change our present picture of the world. The Lord, as Einstein said, is really too subtle! [9]

A sensible reader may immediately notice that the spontaneous motion evidently contradicts the common sense of causality, according to which no event or change can happen without a cause. The common sense is the most basic belief in both daily life and scientific research. It makes the world comprehensible. Imagine you catch a cold and then see a doctor, but the doctor says that your illness has no cause. How surprised you will be! So it is really amazing that motion has no cause. It appears absolutely incomprehensible. Then why do objects move without cause? This is a reasonable question. Only after answering it can the spontaneous motion be comprehensible. Here we leave the suspense to you. We will answer this higher question in the last chapter of this book. There you will find the Lord is subtler. He veils the deeper universal cause of spontaneous motion in the dark. Seeking this ultimate cause is actually a holy road to the Old One.

In the next chapter, we will first answer another easier question: how do objects move if motion has no cause? To our great surprise, we can finally discover the real picture of motion by answering this question.

4
God Plays Dice

I am at all events convinced that He does not play dice.[1]
—Albert Einstein

Motion has no cause. So it must be random. As thus, God does play dice with the universe. Einstein once firmly swore, "In that case, I would rather be a cobbler, or even an employee in a gaming house, than a physicist."[2] But if he knew that there is a logical road to random motion, he would withdraw his remark and gladly be a physicist. Indeed, in order to know God's thoughts, we need logic.

A logical road

The Master arrives without leaving,
sees the light without looking,
achieves without doing a thing.
—Lao Tzu, *Tao Te Ching*

You move everyday, but I can safely say that you don't know how you move. Don't be ashamed for this. In fact, nobody has known the answer of the great puzzle of motion as yet. Fortunately, we have found the crux of the matter: motion involves change in position, while such a change has no cause. This is a crucial clue to discover how on earth objects move. Surprisingly, the answer is very simple.

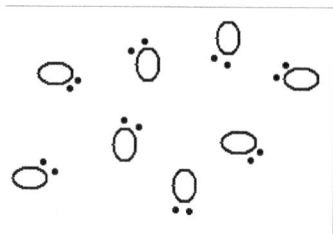

Figure 4.1 I don't know how to move

We first consider a freely moving object. The object is in one position at an instant, and spontaneously appears in another position at next instant. The position of the object is constantly changing, but no cause determines the position change of the object. So the object doesn't "know" how to move at each instant, and can but move in a purely random way. Indeed, nothing causes it to move in a special way. This argument is logical. If a change results from a cause, then the change will be determined by the cause in a lawful way. On the other hand, if a change happens without a cause, then the change must be random. Therefore, the position change of a freely moving object must be essentially random. It should be stressed once again that the object has no velocity to sustain for determining the change of its position between instants. Thus the free object really doesn't "know" which direction it should move along at every instant, and must move in a random way.

The random change of position between instants means that the positions of a moving object are independent of one another. For instance, the object is in one position at an instant, and at the instant immediately neighboring it randomly appears in another position, which is probably not in the neighborhood of the previous position. As thus, the trajectory of the object must be not continuous but discontinuous. Since the change of position is random at all times, the trajectory must be discontinuous everywhere. In this way, the object will always move from one position to another without passing the in-between positions. In a word, the motion of free objects is essentially discontinuous and random.[3]

We then consider the motion of an object influenced by an external force. Can the external force determine the position of the object at every instant and change the random discontinuous motion to deterministic continuous motion? The answer is negative. The reason lies in that a purely random process cannot be changed to a deterministic process in essence. If the external force is not random, then it is evident that the motion of an object under its influence will still be random. If the external force is also random, then since the combination of two random processes still leads to a random process, the motion of an object under such an influence will also be random. So the motion of an object under the influence of an external force is still discontinuous and random.

To sum up, objects move in a discontinuous and random way in reality. During the course of motion, the transition from one position to another is discontinuous and random in nature. In short, God does play dice with the universe. This is a too big surprise. We start only from ordinary phenomena of motion such as the flight of an arrow, and what we use is just simple logic, but where we reach is actually the real picture of motion; the arrow flies in a random and discontinuous way. The picture is so strange that we can hardly believe. But it is real!

In order to make the random discontinuous motion comprehensible, it may be necessary to further answer two relevant questions. One is that how it can accord with our macroscopic experience of continuous motion. If the motion of objects is essentially discontinuous and random, then why does the motion of macroscopic objects appear continuous? The complete answer requires a rather detailed analysis. We will give it in the following chapters. Here we only give an intuitive answer. The crux of the matter lies in that random discontinuous motion usually happens in extremely short space and time intervals for macroscopic objects, which cannot be directly identified by us. As thus, a large number of minute discontinuous motions may generate the average display of continuous motion. In addition, the law of random discontinuous motion will also help to produce an illusion of continuous motion. So the motion of macroscopic objects appears continuous.

The other question is about the displays of random discontinuous motion itself. We haven't seen such strange motion in the macroscopic world after all. If motion is really random and discontinuous, then where can we find it more directly? As we have seen, the double-slit experiment with microscopic particles has clearly revealed that motion is discontinuous. Besides, the randomness of motion can also be found everywhere in the microscopic world. Now let us see two typical evidences.

Evidences

When you insert a straight stick in water, you will surprisedly find that the stick appears bent. Yet, this is in fact an optical illusion resulting from the refraction of light. Nature always hides her secret with attractive veiling. This is also true for the motion of objects. Although everything around us appears to move in a continuous and lawful way, motion is actually discontinuous and random, as some evidences have revealed. The first evidence we will discuss is partial reflection of light.

Partial reflection

> *To see a World in a Grain of Sand*
> *And a Heaven in a Wild Flower,*
> *Hold Infinity in the palm of your hand*
> *And Eternity in an hour.*
> —William Blake, *Auguries of Innocence*

The phenomenon of partial reflection of light is very common in everyday life. Yet, a deep mystery of nature just secretes itself in this simple phenomenon. If we can see it, we can see the whole world.

It is well known that when light falls upon a sheet of usual glass, it is always be partly reflected and partly transmitted. As another beautiful example, when you walk along a lake at a night full of the moon, you can see both the fishes and the moon in the lake. The light from the moon are both transmitted and reflected on the surface of the lake. The transmitted light lets you see the fishes in the lake, while the reflected light lets you see the moon image in the lake.

In the history of science, the partial reflection of light is the first inkling of random motion. This is not unexpected, as today we know light is composed of photons, the most easily accessible microscopic particles. However, even the greatest man in science who observed this phenomenon could not understand it then. The man is Newton, the father of classical mechanics.

Newton tried to explain the behavior of light in terms of particles, called corpuscles. This was convenient for him because then these corpuscles, just like the planets, are subject to the same laws of motion that he had already discovered. So, Newton proposed that light consists of small particles that travel in straight lines. However, Newton's theory of light ran into problems when explaining partial reflection. The puzzle was when a corpuscle of light encounters the surface of a glass, why is it sometimes transmitted and sometimes reflected? How does the corpuscle actually "decide" whether to bounce back or to go through?

To explain partial reflection using Newton's laws of motion, it might be natural to suppose that the distinct structure of reflecting surface causes the unexpected partial reflection. This was also Newton's first attempt to explain the partial reflection of light through glass. He supposed that there are 'holes' and 'spots' in the glass, so that when the corpuscle encounters a hole it goes through and when it strikes a spot it is reflected. Yet Newton himself realized that this theory did not work. He made his own lenses and mirrors by polishing glass. He knew that the small scratches which he made with powder when he polished glass had no observable effect on the partial reflection of light.

Another attempt of Newton to explain partial reflection is a theory of fits. In the case of partial reflection by two or more surfaces, he argued in his *Opticks* (1704) that light striking the first surface excites waves of vibrations that travel along with the light.[4] As a result, the corpuscles of light will undergo a periodic change of state, swinging back and forth between fits of easy reflection and fits of easy transmission. This then leads them to reflect off or pass the second surface. However, it is evident that Newton's theory of fits cannot consistently explain the partial reflection of light by one surface such as the surface of lake.

Today, we know Newton's idea of light consisting of corpuscles is basically valid notably due to Einstein; light is indeed composed of photons. So the puzzle faced by Newton still exists. Fortunately, we have learned more details about the partial reflection of light. We can assure that the input light consists of identical photons, and the structure of

---

Einstein's light

It is Einstein who first revived Newton's idea of light consisting of corpuscles in a deeper level. In 1905, he proposed that light is composed of light quanta in order to account for the photoelectric effect. Different from Newton's situation, however, it was not easy for Einstein to take this step, as it was widely accepted that the wave model of light had been definitively confirmed by experiments then. So Einstein only called his idea of light quanta "very revolutionary". Moreover, Einstein knew more facts of light than Newton. He first realized that light consists of light quanta, and these quanta behave like wave. Yet Einstein stopped his revolutionary step here. He couldn't accept the randomness of motion, which turns out to be the bridge between wave and particle, thus he hadn't understood the peculiar duality of light for life. As he said in old age, "All these fifty years of conscious brooding have brought me no nearer to the answer to the question, 'What are light quanta?'" As a result, Einstein was more plagued by light than Newton.

Interestingly, it seems that light always brought good fortune to Einstein. At the age of 16, he began to chase a beam of light in his mind. This finally leaded him to the famous theory of special relativity. Besides, it was also the deflection of light that helped to confirm general relativity, "the happiest thought" of his life. This might be not unexpected, as light is the little spirit of nature, which is the most accessible but also the most mysterious. If you are enthralled by the beauty of nature, light will also become your fortunate companion. When you see the light, you will see the ultimate reality.

reflecting surface doesn't influence its partial reflection. We can even observe the partial reflection of a single photon, and find that it is indeed sometimes transmitted and sometimes reflected in a random way. Then there is only one possible explanation of the phenomenon, i.e., that the motion of photons is essentially random. As thus, the partial reflection of light is indeed an evidence of the randomness of motion, although the evidence was so faint in Newton's times.

Half life

> *The cause and origin of the radiation… still remain a mystery.*
> —Ernest Rutherford

When physicists began to explore the microscopic world at the end of the nineteenth century, they found the first convincing evidence of the randomness of motion. It was discovered that the atoms in some unusual substances could decay and emit radiation in a spontaneous and random way. Such a phenomenon was called radiative decay or radioactivity. Today it is well known that radioactivity is very dangerous, for example, it can cause the lethal leukemia. Yet it is just radioactivity that let us hear the first clear sound of God playing dice.

The French physicist Henri Becquerel first discovered the phenomenon of radioactivity in a sample of fluorescent uranium salt in 1896. But the term radioactivity was actually coined by Marie Curie, who is probably the greatest woman scientist so far. Inspired by Becquerel's work, Marie further showed that radioactivity is a property of individual atoms of certain species such as thorium in 1898. Two years later, the British physicist Ernest Rutherford observed that emanation from thorium lost half its activity in about a minute, and the intensity of the radiations given out by his sample fell off with time in a geometric progression. He then made the discovery of a half life and the decay law.

The existence of a half life implies that radioactive decays are completely random. Suppose there are 100 million atoms in a radioactive substance at an initial instant. According to the decay law, after a time of the half life of the atoms, 50 million atoms will decay, while the left 50 million atoms will be unchanged. But the decay law cannot tell us which atom will decay and which atom will live longer. So the decay process of a single atom is completely random. The British physicist James Jeans once described this situation in a visual way: "It seemed to remove causality from a large part of our picture of the physical world. If we are told the position and the speed of motion of every one [of a set of radium atoms], we might expect that Laplace's super-mathematician would be able to predict the future of every atom. And so he would if their motion conformed to the classical mechanics. But the new laws merely tell him that one of his atoms is destined to disintegrate today, another tomorrow, and so on. No amount of calculation will tell him which atoms will do this." [5]

Certainly, people can still conjecture that there might exist a deeper law which can determine when the single atom decays in a radioactive substance. However, such a law hasn't been found yet. In fact, more and more evidences have shown that the atoms in a radioactive substance are identical in every aspect. As thus, there is indeed no cause to determine when a specific atom decays. So the decay process must be random in essence.

Today we know the randomness of motion exist everywhere in the microscopic world. In fact, each "click" of the detectors of microscopic particles is just the sound of God playing dice.

A mathematical viewpoint

Logic sometimes makes monsters.
—Henry Poincaré

The existence of random discontinuous motion is also natural and logical from a mathematical point of view. A mathematical proof will undoubtedly give us more confidence in its reality. Fortunately, ordinary readers can also understand this argument.

The motion of an object can be described by a relation of correspondence between each instant and its position at this instant. Such a relation is called function in mathematics. As thus, continuous motion is described by continuous functions, while discontinuous motion is described by discontinuous functions. Our question is: which functions generally exist in mathematics?

This question can also be put in another way. As we know, motion doesn't exist at instants for a moving object; the object is only in one unchanged position at each instant. So the motion state of an object is not its state at an instant, but its state during a finite interval or an infinitesimal interval near the instant. Since a time interval contains infinitely many instants, the motion state of an object will be described by a set of points in space and time, in which each point represents the position of the object at each instant. Then what is the general form of such a point set? Is it a continuous line? Or is it a discontinuous point set? The former corresponds to continuous motion, while the latter corresponds to discontinuous motion.

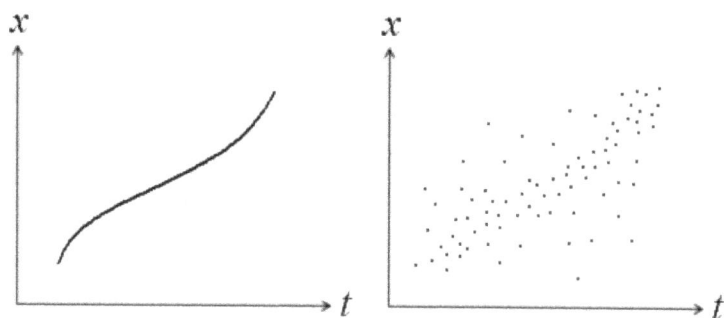

Figure 4.2 Continuous motion and discontinuous motion

Mathematicians had answered the above questions at the beginning of the twentieth century. Before then mathematicians were only familiar with continuous functions. Their existence accords with everyday experience. Especially motion is apparently continuous, and thus can be directly described by continuous functions. However, mathematics began to rely more on logic rather than on experience in the end of the nineteenth century. Many discontinuous functions were then discovered in the mathematical world, one of which is that its value is zero at rational points and is one at irrational points.

At first, most orthodox mathematicians were very hostile to the discontinuous functions. They called them pathological functions. As the French mathematician Henry Poincaré remarked in 1899, "Logic sometimes makes monsters. For half a century we have seen a mass of bizarre functions which appear to be forced to resemble as little as possible honest functions which serve some purpose. More of continuity, or less of continuity, more derivatives, and so forth... In former times when one invented a new function it was for a practical purpose; today one invents them purposely to show up defects in the reasoning of our fathers... It is the beginner that would have to be set grappling with this teratologic museum."[6]

However, some young French mathematicians, notably Emile Borel and Henri Lebesgue, took them very seriously. They discovered that discontinuous functions could be strictly analyzed by applying the theory of sets. As thus, they led a revolution in mathematical analysis.[7] The core notion of this revolution is measure. Simply speaking, measure is an extension of length. Length only applies to describe continuous lines, which are special point sets, while measure is a universal descriptive quantity of all point sets. As an example, the set of all rational points between 0 and 1 in a real line is a dense point set, and its measure is zero. Certainly, the measure of a line still equals to its length.

According to the measure theory, continuous functions are very special functions. Nearly all functions are actually discontinuous functions. As Poincaré admitted, "Indeed, from the point of view of logic, these strange functions are the most general... If logic were the sole guide of the teacher, it would be necessary to begin with the most general functions." [8] In the language of measure, the measure of the set of all continuous functions is zero, while the measure of the set of all discontinuous functions is one. In this sense, the existing probability of continuous functions is zero in the mathematical world.

On the other hand, according to the point set theory, general point sets are random discontinuous point sets. By comparison, continuous lines are very special point sets. The measure of the set of all continuous lines is zero. This conclusion is logical and can also be understood intuitively. Imagine you draw in dots on a paper at will. It must be extremely improbable that these dots can be connected to form a smooth line.

So, from a mathematical point of view, continuous motion, which is described by continuous functions and continuous lines, cannot exist, while random discontinuous motion, which is described by discontinuous functions and random discontinuous point set, must exist.

In a word, we find that random discontinuous motion has a solid position in the mathematical world. If the great book of nature is indeed written in the language of mathematics, as Galileo put it, then motion must be discontinuous and random in nature.

Two seers

> *Man is but a reed, the most feeble thing in nature, but he is a thinking reed.*
> —Blaise Pascal

Although random discontinuous motion is so strange and even mind-boggling, two ancient thinkers had ever proposed similar ideas for different purposes.

The first man who seriously considered the randomness of motion is the Greek philosopher Epicurus, who lived in the same times with Aristotle. He presented the well-known idea of atomic "swerve" on the basis of Democritus's atomic theory.[9] Like Democritus, Epicurus also held that the elementary constituents of the world are atoms, which are indivisible microscopic bits of matter, moving in empty space. He, however, modified Democritus's strict determinism of elementary processes. Epicurus thought that occasionally the atoms swerve from their course at random times and place. Such swerves are uncaused motions. One reason for this swerve is that it is needed to explain why there are atomic collisions. According to the atomic theory, the natural tendency of atoms is to fall straight downward at uniform velocity. If this were the only natural atomic motion, the atoms would never have collided with one another, forming macroscopic bodies. Therefore, Epicurus saw it necessary to introduce the random atomic swerves.

The second man who clearly presented the idea of discontinuous motion is the Arabic theologian Abu Ishaq Ibrahim Al-Nazzam (d. 845).[10] He called it the theory of "leap". In his treatise on motion, *Kitāb fial-haraka*, Al-Nazzam attempted to solve Zeno's paradoxes by proposing that a body could move from a position to another in space without passing the in-between positions. He argued that an object in motion performs a leap: "The mobile may occupy a certain place and then proceed to the third place without passing through the intermediate second place on the fashion of a leap."[11] Moreover, a leap from position A to position B consists of two interlocking sub-events; the original body in position A ceases to exist, and an "identical" body comes into being in position B. As thus, Al-Nazzam's leap theory reduced the apparently continuous motion of macroscopic objects to a sequence of microscopic discontinuous processes.

It can be seen that neither Epicurus's atomic swerve nor Al-Nazzam's leap motion is exactly the random discontinuous motion; the former is random but without explicit discontinuity, while the latter is discontinuous but without explicit randomness. Certainly, the latter is more like it than the former concerning the picture of motion. In a sense, random discontinuous motion can be regarded as an integration of Epicurus's random swerve and Al-Nazzam's discontinuous leap. In the next chapter, we will look at it more closely.[12]

5
Time Division Universe

There is time for everything.
—Thomas Edison

Subtle is the Lord! He does play dice with the universe; motion is actually random and discontinuous in nature. This revelation will open a wholly new world before us. It is unimaginable but real. It is strange but comprehensible. You can enjoy the amazing world now.

Here we will give a clear picture of God playing dice in the atomic world. In the picture, a particle is no longer a local particle, but a nonlocal particle cloud. Moreover, different branches of a particle cloud can superpose and "interfere" with each other like a wave. More surprisingly, two particle clouds can be entangled to form an inseparable whole in a time division form, and the wholeness is not impaired at all no matter how far they are separated. No doubt the new picture will lead to a profound shift in our world view. It implies that the universe is not a mere aggregate of independent existences, but an inseparable whole in the time division form. We live in a time division universe in reality.

As a byproduct, we will also find that the most mysterious quantum language, which was originally invented by the Austrian physicist Erwin

Schrödinger more than eighty years ago, is just a delicate description of the particle clouds and their evolution. Now laymen as well as physicists can finally understand it.

A particle is a lone cloud

> *Exactly what is an electron?*
> —Richard Feynman

The motion of a particle is essentially discontinuous and random. From the point of view of the particle, it has a propensity to be in any possible position at each instant. The particle is in one position at an instant, but at the instant immediately neighboring it randomly appears in another position, which is probably not in the neighborhood of the previous position. In this way, the particle jumps from one position to another without passing through the in-between positions.

This is a picture of motion in terms of instants. In the picture, the particle is restless at all times like a living body. It wanders about, guided only by will. The particle, which is usually regarded as an inert object, is actually alive! It is really a leaping spirit. If you close your eyes and imagine this picture for a moment, you will feel the amazing activity of an ordinary particle. It may change your picture of the world forever. But this is just a beginning.

Let us now open time window, and look at the motion picture of the particle during a time interval. In this picture, although the random activity of the particle is concealed, it will have other more amazing displays. The picture is pivotal for our deep understandings of the motion of particles and its law.

The particle is only in one position at an instant, while during a time interval, which contains infinitely many instants, its positions will form a set of points in space and time, in which each point represents the position of the particle at each instant. Due to the essential discontinuity of motion, the point set is discontinuous and generally spreads throughout the whole space. Obviously, the point density of the point set in each position represents the frequency of the particle appearing in that position. The more frequently the particle appears in a position, the larger the point density is at that position.

If the time interval is very small or even infinitesimal, the above discontinuous point set will accurately represent the motion state of a particle near an instant. It is thanks to Lebesgue that the point set can be strictly described in math using his measure theory. Its complete description includes position measure density and position measure flux density.[1] The former describes the point density distribution of the point set, and the later describes the change of this density distribution with time.

The intuitional picture of the motion of a particle is that the point set representing its motion state spreads in space like a cloud. For a brief and graphic description, the point set will be called particle cloud.[2] We will discuss the motion of particles in the new language of particle cloud from now on. A particle is no longer a local particle, but a nonlocal particle cloud concerning its picture during a time interval. The cloud is generated by the random discontinuous motion of a particle during an infinitesimal time interval near an instant, and each point in the cloud represents the position of the particle at each instant. As thus, a particle cloud can visually represent the motion state of a particle. Especially, the density of the particle cloud in a position just represents the frequency of the particle appearing in the position.[3] The particle cloud is denser in the region where the particle appears more frequently (e.g. at the center of the figure).

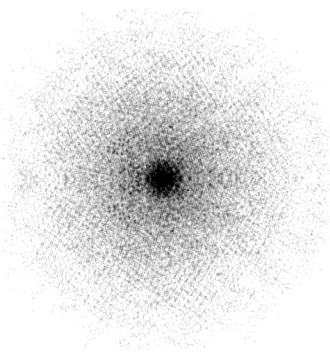

Figure 5.1 A particle cloud

Let us first see the simplest particle clouds. We may call them basic particle clouds or basic clouds. A basic cloud has an even density distribution in the whole space. This means that the particle has the same propensity to be in every position in space, and thus the frequency of the particle appearing in every position is the same. So, a basic cloud is indeed the most natural and simplest motion state of a particle.[4]

A basic cloud can have a non-zero velocity, and its velocity can be changed by an external force or interaction. This is just like the situation where force causes the change of the velocity of a classical particle in Newton's imaginary world of continuous motion. As to the actual discontinuous motion of a particle, the particle has no continuous trajectory and thus has no velocity, but velocity can be well defined for the particle cloud.

For a basic cloud with a non-zero velocity, its density is still constant throughout the whole space, but the whole cloud (*not* the particle) will move with a definite velocity. Since the motion of the basic cloud doesn't result in any change of its density distribution, the cloud appears motionless. This is very similar to the situation of steady current; the current appears quiet, but if you put your feet in the current, you will feel its motion. In fact, every local part of the particle cloud moves with the same velocity as the whole cloud. In physics, the motion of a basic cloud is also described by another quantity called momentum, which is defined as the mass of the particle multiplied by the velocity of the basic cloud.

Next, let us see the general particle clouds. A general particle cloud has uneven density distribution, and the local parts of the cloud usually move with different velocities. The local motion can be described by flux density, which equals to the product of density and its local velocity.[5] Density distribution and flux density distribution provide a complete description of a particle cloud. This is similar to the description of a fluid in hydromechanics. Since the different parts of a particle cloud usually have different velocities, the particle cloud will diffuse during its whole motion.

Although the size of a particle is very tiny, the size of a particle cloud can be as large as that of a ball in actual situations. For instance, a photon cloud from a distant star is like a very wide, bubble-thin disk. Its width can vary from a few feet to several milimeters in diameter though its thickness is smaller than a soap bubble. By comparison, the width of a photon cloud from the sun is only a small fraction of a millimeter. When such a photon cloud falls upon a sheet of glass, it will be divided into two branches in general, one of which is reflected and the other is transmitted. The photon then jumps between these two separated branches in a random way. This is the actual process of the partial reflection of light, which had ever puzzled Newton.

Besides, a particle cloud can readily pass through two slits at the same time when the distance between the slits is smaller than the width of the particle cloud. More than 200 years ago, many lucky photon clouds from the sun passed through the two slits set up by the British scientist Thomas Young.[6] They helped to accomplish one of the most beautiful experiments in the history of science, namely Young's double-slit experiment, which definitively proved that light has wave properties. We will explain why a particle cloud behaves like a wave in the next section.

Lastly, let us enjoy some beautiful still particle clouds. For the imaginary continuous motion of a particle, there always exists a rest state when selecting a suitable frame of reference. But as to the actual motion of a particle, there exists no rest state for the particle in nature, as the particle jumps in a random and discontinuous way at all times. However, there are still rest states for its motion. They are the motion states of the particle that don't change with time. Concretely speaking, the particle cloud is at rest in space, and its density distribution doesn't change with time. Such states are often called stationary states. Since a particle cloud tends to diffuse during its motion, a stationary state usually needs to be bounded by an external force. This is essentially different from the situation in Newton's classical world.

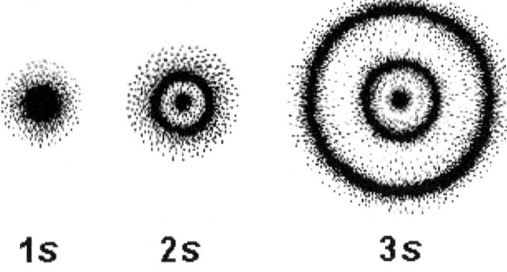

Figure 5.2 Electron clouds in a hydrogen atom

Modern life depends on electrons. We use electricity for lighting, cooking, and nearly every doing. But do you know exactly what an electron is? An electron is actually an electron cloud, which is formed by its random jump motion, and it can be in a stable stationary state in an atom. Please enjoy the beautiful electron clouds in a hydrogen atom. Note that their size is no more than a few billionths of an inch.

As a familiar example, the actual motion of an electron will generate a still electron cloud in a hydrogen atom, as the electron is attracted by a proton there. The electron cloud can have different density patterns, which correspond to different discrete energies of the electron. The density of an electron cloud in each position represents the frequency of the particle appearing in that position. The electron cloud is denser in the region where the electron appears more frequently. Note that the appellation "electron cloud" has been used in textbooks by physicists and chemists, but with different meaning. It is only an imaginary probability cloud there, which density is proportional to the probability density for the electron being found around the atomic or molecular nucleus. Certainly, if you are familiar with the probability cloud, you will understand the real picture of electron cloud more easily, as they have the same patterns.

In a word, a particle is not a local particle, but a nonlocal particle cloud in the atomic world. Every thing (i.e. entity existing in space and time) is actually a thing cloud. This will lead to a profound shift in our picture of the world. Next time when you are asked exactly what an electron is, you may answer poetically that an electron is a lone cloud.

The particle cloud ripples

The one great dilemma nails us... day and night is the wave-particle dilemma.
—Erwin Schrödinger

Throw a stone into a still lake and watch the ripples spread in rings. How beautiful nature is! To our surprise, a particle cloud will also ripple when it is disturbed. This will help people finally get out of the great wave-particle dilemma, which had plagued many realistic physicists including Einstein and Schrödinger night and day. Why does a particle behave like a wave? Because a particle is actually a particle cloud, and the cloud behaves like a wave.[7]

We take a basic cloud as an illustration. Suppose a basic cloud is reflected by a wall. The reflection disturbs the basic cloud. The whole cloud is divided into two branches after reflection: the input branch and the reflected branch, which have the same density and move with opposite velocities. They then superpose in the same space to form a new particle cloud.

How about the new superposed particle cloud? At first glance, it seems that its density should be a direct addition of the densities of the two branches. Since the density of each branch is constant in space, the density of the superposed particle cloud will also be constant in space. However, this commonsense view turns out to be wrong; two branches of a basic cloud can actually superpose and "interfere" with each other just as two waves superpose and interfere, and the density of the superposed particle cloud is not constant but periodic in space. This consequence seems very counterintuitive, but it is logical and true.

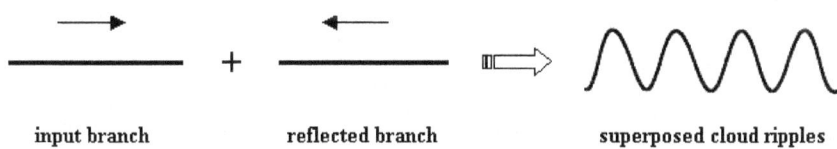

input branch   reflected branch   superposed cloud ripples

Figure 5.3 Two branches of a basic cloud superpose and ripple

Let us find the density distribution of the superposed particle cloud. Before the superposition, there is a particle cloud containing two branches, which have opposite velocities. At some instants, the particle is in the input branch, which moves towards the positive direction with a speed, while at other instants the particle is in the reflected branch, which moves towards the negative direction with the same speed. After the two branches superpose in the same space, the superposed particle cloud will be motionless, as the two branches have the same density and move with exactly opposite velocities. This means that the flux density of the superposed particle cloud is zero everywhere.

Then where does the information of the speed go? Such information belongs to the whole particle cloud, and cannot be erased by the superposing process. It must exist in the superposed particle cloud, but it doesn't exist in its flux density distribution, which is simply zero. Since density distribution and flux density distribution provide a complete description of a particle cloud, such information must exist in the density distribution of the particle cloud. As thus, the superposed particle cloud cannot have a constant density in space, which can assume one in every position, but must have an uneven density distribution relating to the speed. This is a consequence of logic. A detailed mathematical analysis further shows that the density distribution is periodic in space, and the period is inversely proportional to the speed.[8]

The result is a big surprise. It is against all expectations. Two branches of a basic cloud with even density distributions can actually superpose to form a new particle cloud with uneven but period density distribution. This is very similar to the situation where two traveling waves are superposed to form a standing wave; the superposed particle cloud is really like a standing wave. Thus, a basic cloud indeed behaves like a wave. No doubt this result has implied that a particle cloud has an amazing wave-like characteristic.

It can be seen that the wave-like characteristic of an actual particle essentially results from the discontinuity of its motion. If motion is continuous, the superposition of two velocities will be a direct addition of them. So no wave-like characteristic exists for a classical particle undergoing continuous motion. In fact, continuous motion is local in nature. For instance, a particle moving continuously cannot pass through two slits like a wave at all. By comparison, the particle cloud generated by the discontinuous motion of a particle can readily pass through two slits at the same time, and its two divided branches can then superpose and "interfere" with each other just as two waves superpose and interfere. This is the very origin of the double-slit interference pattern.

Two fabrics

> *When yin and yang combine, all things achieve harmony.*
> —Lao Tzu

We have shown that two branches of a basic cloud with opposite velocities can superpose and "interfere" with each other like a wave. But this is only a special illustration. Does every particle cloud behave like a wave? In other words, is the wave-like characteristic a general characteristic of particle clouds? In order to answer this question, we need to further probe the fabric of particle clouds with a logical microscope.

In a particle cloud, the particle stays in one position at an instant, but during an infinitesimal time interval the particle will jump throughout the space spread by the particle cloud. As a result, locality and nonlocality coexist in a particle cloud. Correspondingly, there are two kinds of basic clouds: local basic clouds and nonlocal basic clouds. The nonlocal basic clouds are just the basic clouds discussed above. They have one determinate velocity or momentum and spread throughout the whole space with an even density distribution. The local basic clouds are the particle clouds which are concentrated in one determinate position. For a convenient expression which is consistent with the existing appellation, the local basic clouds and the nonlocal basic clouds will be called position bases and momentum bases.[9]

Every particle cloud can be formed by both a certain superposition of position bases and a certain superposition of momentum bases. This means that a particle cloud has two fabrics: one (local) position fabric, the other (nonlocal) momentum fabric, and there exists a one-to-one relation between them.[10] In fact, as to every particle cloud in real (position) space, we can imagine that in the abstract momentum space there is a corresponding particle cloud, in which one point represents one momentum in the superposition. They respectively represent the position fabric and the momentum fabric of the particle cloud, and there is a one-to-one relation between them. In math, a proper integrative description can be constructed in terms of density distribution and flux density distribution for the particle clouds in both spaces, and the one-to-one relation can then be represented by a transformation between the integrated position description and the integrated momentum description.

As we will see, the transformation is the key to reveal the general wave-like characteristic of particle clouds. Then how to find it? To our surprise, symmetry can lead us to it. As to a momentum basis, its particle cloud in position space has an even density distribution throughout the whole space. This means that a momentum basis is composed of all position bases. Similarly, as to a position basis, its particle cloud in momentum space has an even density distribution throughout the whole momentum space, and thus a position basis is also composed of all momentum bases. This implies that position and momentum are essentially symmetrical for describing the particle clouds. As a consequence, the transformation between position description and momentum description will have corresponding symmetry. Concretely speaking, the transformation, as a function of position and momentum, is invariant under the exchange of position and momentum, and the reverse transformation is the same as the transformation in which the sign of position or momentum is reversed.[11]

From a general point of view, the above symmetry essentially results from the distinct character of random discontinuous motion, especially the dialectic relation between locality and nonlocality coexisting in it. They not only are opposite each other, but also embody one another. For instance, a nonlocal (momentum) basis is composed of all local (position) bases, while a local (position) basis is composed of all nonlocal (momentum) bases. Locality and nonlocality, which can be represented by quiet staying and restless jumping respectively, are very like yin and yang in Chinese philosophy, and they are unified as a perfect whole in the random discontinuous motion. As we will see later, the inherent symmetry and harmony of motion will essentially determine the law of motion.

Yin and Yang are two primal opposing but complementary forces existing in all things in the universe. Each contains the seed of the other. Their constant interaction enables change to take place within the world.

Figure 5.4 Chinese Yin-Yang Diagram

When considering the finiteness requirement in physics, the symmetry of the above transformation may uniquely determine its form in math. It turns out that the transformation is the well-known Fourier transformation, in which position and momentum are the paired variables.[12] In the Fourier transformation, a dimensional constant must be introduced to cancel out the dimension of the product of position and momentum. Historically, the German physicist Max Planck first introduced this constant for other reason in 1900, which was called Planck constant from then on. Here we can see its real physical origin.

An interesting property of Fourier transformation is that the multiple of the dispersions of the two variables in the transformation has an upper limit. In the above Fourier transformation, the two variables are position and momentum, and the upper limit equals to Planck constant divided by 4π. Historically, this relation between position and momentum was first discovered by the Germen physicist Werner Heisenberg in 1927, and was later called Heisenberg's uncertainty principle.[13] Here we can see that it is just one of the properties of random discontinuous motion.

As we know, the Fourier transformation, which was first introduced by the French mathematician and physicist Joseph Fourier early in the nineteenth century, is used to analyze periodic phenomena and wave motion – the vibrations of a violin string, the oscillations of a clock pendulum etc. So the above result has implied that the wave-like characteristic is a general characteristic of particle clouds. Then which form of wave is it? How does this wave evolve?

Psi wave and its evolution

Erwin with his psi can do
Calculations quite a few.
But one thing has not been seen
Just what does psi really mean.
—Walter Hückel translated by Felix Bloch

In order to see what wave a particle cloud is and how it evolves, we first look at the simplest momentum bases. By using the Fourier transformation between position description and momentum description, we can immediately find the integrated position description of a momentum basis of a particle. It turns out to be a complex wave function describing a plane wave at a given instant, which wavelength equals to Planck constant divided by the momentum of the particle.

The time-included form of the wave function can be further determined by the meaning of momentum.[14] In this complete form, a new quantity, which is paired with time, will appear. This quantity is called energy today. For the situation where the mass of particle is a constant quantity, it equals to the square of momentum divided by two times of the mass of particle.[15] The wave function describes a plane wave propagating in space. Its period equals to Planck constant divided by the energy of the particle.

The (time-included) wave function of a momentum basis has actually determined its free evolution. The free evolution equation proves to be a linear wave equation. Owing to the linearity of the equation, it also holds true for the linear superpositions of momentum bases, which form the wave functions describing all particles clouds due to the completeness of the momentum bases. So this equation is also the free evolution equation of the wave function describing any particle cloud.

By using the free evolution equation, we can further find how the wave function is integrated from the density and the flux density of a particle cloud. The amplitude of the wave function equals to the square root of density. The phase of the wave function relates to the space integration of flux density divided by density (i.e. of the local velocity of the particle cloud). As thus, the wave function, which is an exquisite mathematical complex of density and flux density, also provides a complete description of the particle cloud generated by the random discontinuous motion of a particle.

In practical situation, the wave function is more convenient than density and flux density due to its linear superposition. We take the double-slit experiment as an illustration. In the experiment, a particle cloud passes through two slits and is then divided into two branches. But the two branches, especially their densities, don't directly superpose in the space between slits and screen. What superpose are the two wave functions describing the branches. Since these two wave functions generally have different amplitudes and phases in space, they will interfere just like two water waves. Then the amplitude of the superposed wave function will have a wave-like interference pattern in space, so does the density of the superposed particle cloud, which equals to the square of the amplitude of the wave function. This provides an intelligible explanation of the double-slit experiment in terms of the wave function.

Lastly, let us consider the evolution of the wave function under an external potential. Since potential is a classical description of interaction, we can use Newton's second law of motion as a limit.[16] It proves that the evolution equation of the wave function under an external potential takes the same form of the Schrödinger equation in quantum mechanics.[17] In this way, Schrödinger's equation is rediscovered along a logical road.

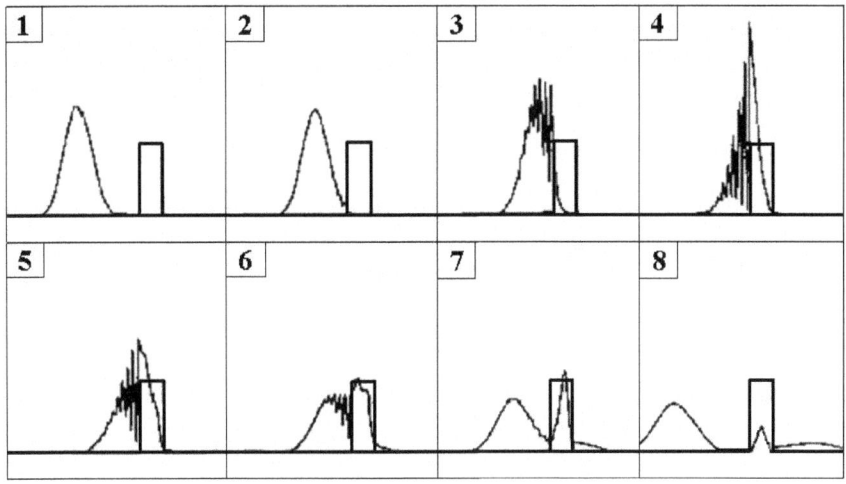

Figure 5.5 The visual evolution of a particle cloud

A particle cloud (depicted as a wave packet) moves towards a potential barrier, which is like a wall. Its evolution follows the Schrödinger equation. Most parts of the particle cloud are reflected, but a small part of the particle cloud tunnels through the wall[18]. This tunneling process is usually called quantum tunneling. It is the basis for transistors, which are widely used in radios, televisions, computers, and so on.

During the Christmas holidays of 1925, Schrödinger invented his famous wave equation to account for the atomic phenomena at the alpine resort of Arosa, Switzerland. He represented the wave function in the equation with psi, the 23rd letter of the Greek alphabet. Einstein remarked immediately in a letter to Schrödinger, "The idea of your work springs from true genius!"[19] To our surprise, however, Schrödinger didn't know what his equation describes, as he reached it not by logic but by analogy and guess. More surprisingly, the Schrödinger equation proved to be awfully successful in application. Today, every physicist uses psi wave and its equation, but nobody really understands them.[20] Could you imagine this absurd situation? Scientists actually talk with nature with a language they don't understand at all! All these will be radically changed after the discovery of random discontinuous motion.

Now we finally know the mysterious psi wave describes the particle cloud generated by the actual motion of a particle, which is random and discontinuous in nature, and the Schrödinger equation is just its evolution law, namely the law of motion.

The universe is an inseparable whole

You are not alone
For I am here with you
Though you're far away
—Michael Jackson, *You Are Not Alone*

Wave-like characteristic is indeed an amazing characteristic of random discontinuous motion; a particle is actually like a wave. However, the deepest implication of random discontinuous motion lies in that it entangles the universe into an inseparable whole in a form of time division. As we will see, when one particle meets another, they immediately become two inseparable lovers.

Let us take two entangled electrons as an illustration. The *pas de deux* is absolutely miraculous. Suppose two electrons are initially independent. Electron 1 is in a stationary state, and its particle cloud has two branches *A* and *B* which evenly distribute in two isolated regions. This means that electron 1 jumps between the two regions in a discontinuous and random way. The particle cloud of electron 2 spreads straight towards the middle of the two regions from the bottom up. There exists a repulsive force between the two electrons, which is well known as Coulomb force. When electron 2 approaches electron 1, they begin to interact with each other observably.

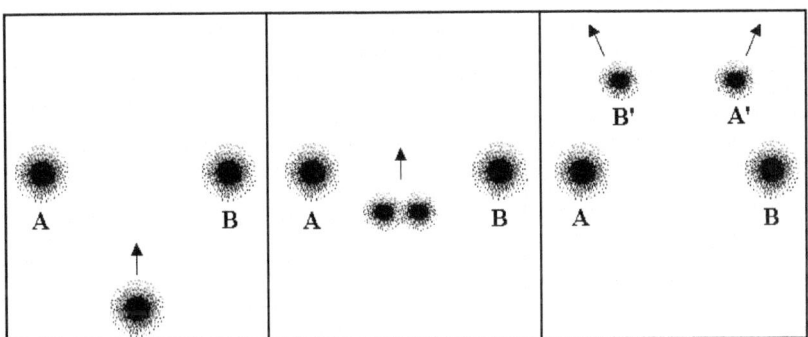

Figure 5.6 The forming process of two entangled electrons

Due to the discontinuity of motion, the two electrons will form a new whole. The forming process is as follows. When electron 1 is in the left branch *A* at some instants, it will cause electron 2 to move rightward by repulsive force. In a similar way, when electron 1 is in the right branch *B* at other instants, it will cause electron 2 to move leftward by repulsive force. Then after electron 2 passes through the regions of electron 1, its particle cloud will be split into two isolated branches A' and B'. Here an inseparable whole is formed, in which the branches A' and B' of electron 2 and the branches A and B of electron 1 are entangled. At any time, when electron 1 is in branch A, electron 2 must be in branch A', while when electron 1 jumps to branch B, electron 2 must also jump to branch B' synchronously. In this way, these two electron clouds are correlated and entangled in the tightest way.

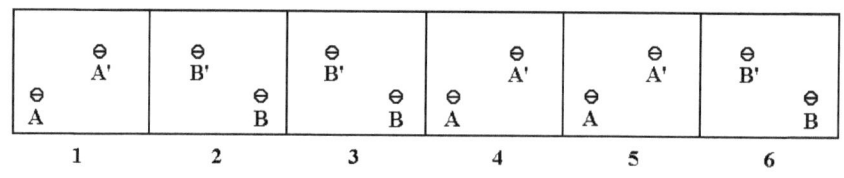

Figure 5.7 Two entangled electrons at six neighboring instants

The whole of the two entangled electrons exists in a time division form. At some instants (e.g. 1, 4, 5 in the above figure), electrons 1 and 2 are in the branches A and A'. These instants constitute one discontinuous dense set of instants. At others instants (e.g. 2, 3, 6 in the above figure), electrons 1 and 2 are in the branches B and B'. These instants constitute the other discontinuous dense set of instants. The two dense instant sets, which can be called time sub-flows, constitute a whole continuous time flow. It looks as if the world is time-divided into many sub-worlds, each one of which occupies one tiny part of the continuous time flow, and the occupying way is discontinuous and random in essence.

5 – TIME DIVISION UNIVERSE

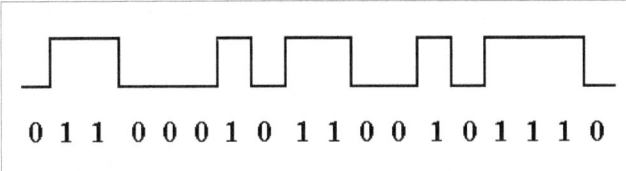

Figure 5.8 Time division multiplexing

The word "time division" may be not strange for some people. One can find many related words in a modern dictionary, for instance, time division telegraph, time division multiplexing (TDM), and time division multiple access (TDMA) etc. In fact, time division is widely used as an effective method to transmit information in modern communication technology; every bit of information on the Internet as well as every voice over the telephone are both coded and transmitted in a time division form. Certainly, the coded information is time-divided in a determinate way, or else we cannot decode it. Now, to our great surprise, our universe is also time-divided, but in a purely random way. Everything, from an atom to a star, lives in the time division universe.

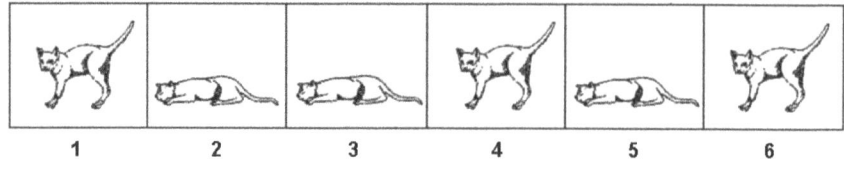

Figure 5.9 The picture of Schrödinger's cat

When a decaying atom interacts with a cat via a set of devices including a detector, a hammer, and a small flask of hydrocyanic acid etc, the whole system will be in an entangled state containing two branches. In one branch the atom decays and poisons the cat, while in the other branch the atom doesn't decay and the cat still lives. As a consequence, the cat will be in an uncertain superposition of dead state and living state. This cat is well known as Schrödinger's cat today. In 1935, Schrödinger first conceived this famous Gedanken experiment. The figure depicts Schrödinger's cat at six neighboring instants in the time division universe. At each instant the cat is either alive or dead in a purely random way.

By extending the description of a particle cloud, we can use a joint position density to describe the two entangled particle clouds. It represents the appearing frequency of the situation in which electron 1 is in one position and particle 2 is in another position. In a similar way, we can further define a joint position flux density. It is worth noting that these two functions don't exist in the three-dimensional real space, but exist in a six-dimensional abstract space. Correspondingly, the integrated wave function of the two entangled particles also "lives" in the six-dimensional space.

The existence of the entangled states of particles (including multi-particle entangled states) has been confirmed by experiment. They are usually called quantum entangled states. It is Schrödinger who first seriously studied such states and coined the term 'entanglement' to describe them in 1935.[21] Today there exist many effective methods to generate the entangled states, one of which is spontaneous parametric down-conversion (SPDC). The SPDC method is widely used to generate the entangled photons such as polarization-entangled photon pairs.[22]

Lastly, we will look at more closely the most amazing characteristic of entangled states, their inseparable wholeness. It seems that there exists one kind of mysterious synchronism between the two electrons in an entangled state. The double dance is absolutely in step. This kind of synchronism is not only accurate in time, but also irrelevant to the distance between the electrons – even if they are at opposite ends of the universe. Moreover, electrons 1 and 2 being in the branches A, A' or B, B' are purely random at any instant due to the inherent randomness of motion. This kind of stochastic synchronism is more mysterious. Then how can the two electrons sustain the stochastic synchronism without any error? Or putting in an anthropopathic way, how can each electron instantaneously "feel" the random change of the other? This phenomenon is absolutely curious!

It is understandable that two lawful processes or two events which have a common cause can have absolute synchronism. Let us take an illustration. Suppose two lights are controlled by one switch, and the distance between the switch and each light is the same. Then the two lights can be turned on or turned off in an absolutely synchronous way.

However, if two processes not only are random, but also have no common cause, then common sense tells us that they cannot obtain synchronism at all. Indeed, if we only use the motion picture of parts to understand the magical random synchronism, we can never find a satisfying answer, as the independent motion of each part doesn't exist at all.[23] The two entangled electrons are actually an inseparable whole in a time division form concerning their motion. The random synchronism between the electrons is just a characteristic of this whole. It is formed and sustained by the time division property of motion.

This consequence will lead to a profound change in our picture of the universe; the universe is not a mere aggregate of independent existences, but an inseparable whole in the time division form. The form of time division, which results from the discontinuity of motion, results in the inseparable wholeness of the universe. In this meaning, random discontinuous motion is the most powerful glue that holds the universe together. We usually view the world in the light of the concept of part. In fact, we should view the world in the light of the concept of wholeness. Reality is a whole in the form of time division.

It is worth noting that no interaction is required to sustain the existence of a whole in the form of time division after it is formed.

Especially, no matter how far the parts are separated, this kind of wholeness is not impaired. This characteristic is essentially different from that of an ordinary whole in everyday life, which needs to be held by interaction.

In a word, all things in the universe are entangled into an inseparable whole in the time division form because of the inherent discontinuity of motion. This is an amazing picture of reality. We actually live in an inseparable time division universe. Yet it seems that the macroscopic world is separable; we seldom experience inseparable wholeness, and we never see Schrödinger's cat either.[24] Why? A deeper secret of nature is still waiting for us.

6
The Gambling Rule

> *Was the wavefunction of the world waiting to jump for thousands of millions of years until a single-celled living creature appeared? Or did it have to wait a little longer, for some better qualified system ... with a Ph.D.? ... Do we not have jumping then all the time?* [1]
> —John Bell

Motion is essentially discontinuous and random. But, up to now, we see no randomness and discontinuity in the particle cloud and its evolution law; the wave function is continuous, and the Schrödinger equation is also continuous and deterministic. Then where do the randomness and discontinuity go? Remember they do appear in the actual experiments on microscopic particles.[2] On the other hand, if motion is really discontinuous and random, then why does the motion of macroscopic objects such as a ball appear continuous and deterministic? Again: where do the randomness and discontinuity go?

These two puzzles are actually connected with each other. The solution of them will reveal a deeper secret of nature; space and time are not continuous but discrete. The discreteness will naturally release the inherent randomness and discontinuity of motion. As thus, God, the cosmic gambler, plays observable dice in discrete space and time.

Moreover, His gambling rule will create the illusion of continuous motion around us. So, random discontinuous motion in discrete space and time will provide a unified picture of the microscopic and macroscopic worlds. It is *the* real motion.

Where does the randomness go?

> *Zeus made it rain continually.*[3]
> —Homer

Although motion is discontinuous and random between instants, the discontinuity and randomness are absorbed into the motion state defined during an infinitesimal time interval in continuous space and time. As a consequence, the wave function as well as density and flux density, which describe the motion state, are continuous. Moreover, the evolution equation of the motion state, namely the Schrödinger equation, is also continuous and deterministic. In short, God plays dice in continuous space and time, but the dice is unobservable in nature. Then where does the randomness of motion go? How can the random motion present itself?

The sticking point lies in the assumed continuity of space and time. As we have pointed out in the first chapter of this book, the continuity of space and time cannot be confirmed in essence; we can never measure infinitely small space and time regions, let alone durationless instants and positions. In more objective words, a durationless instant cannot present itself. So, it is natural that the randomness of motion, which exists at individual durationless instants in continuous space and time, cannot emerge through observable physical effects. We can only measure finite time intervals, but during which there are no randomness at all.

To sum up, if space and time are continuous, then the inherent randomness of motion cannot be released in essence. This result not only contradicts experience, but also is very unnatural in logic. To begin with, there are convincing evidences of randomness in the experiments on microscopic particles; each "click" of the detectors of microscopic particles is the very sound of God playing observable dice. Certainly, we can use other assumed sources of randomness to account for the existing experience. But if they are essentially random and discontinuous, they will also be unobservable in continuous space and time. Besides, assuming two different kinds of fundamental randomness may not satisfy the requirement of Occam's razor. By comparison, it is more natural and simpler to assume that the inherent randomness of motion can emerge and generate the actual randomness of motion.

Next, the above result evidently contradicts one of our most basic scientific beliefs, the minimum ontology. According to the principle, existence should present itself. If a certain thing does exist, then it can be detected, whereas if a certain thing cannot be detected in essence, then it does not exist. So, if the minimum ontology is true, then space and time cannot be continuous, and the inherent randomness of motion must emerge and further generate the actual random phenomena.

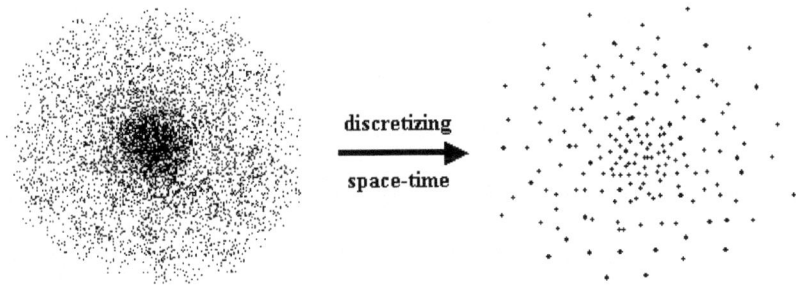

Figure 6.1 Randomness appears in discrete space and time

The continuity of space and time has embedded a suicidal seed in itself, while the random discontinuous motion finally annihilates it. Then what is the actual form of space and time? It must be discrete in nature.[4] Instants and positions are not 0-sized but finite-sized. Especially, time is composed of instants with finite duration, which are the smallest parts of time. Since instants as finite intervals can have physical effects and be measured in principle, the inherent randomness of motion, which exists at instants, may emerge in discrete space and time. Concretely speaking, in a particle cloud, the particle stochastically stays in a position during a finite instant, and the random finite stay may have a tiny effect on the continuous evolution of the particle cloud. Then during a much longer time interval, such tiny random effects may continually accumulate to generate the observable random phenomena.

In a word, the inherent randomness of motion can only be released in discrete space and time. Then does the discreteness of space and time actually release the randomness of motion? Does God really play observable dice in discrete space and time? Is there new surprise brought by the discreteness of space and time? The answers are positive.

Playing dice in discrete space and time

Time and chance reveal all secrets.
—An old adage

We have been discussing the motion of objects in continuous space and time. In this section we will look at the motion of objects in discrete space and time. As we will see, the discreteness of space and time is a very strong restriction. It will bring us many new surprises which are absent in continuous space and time. Especially, it can indeed release the inherent randomness of motion. So, God really plays the observable dice in discrete space and time.

In discrete space and time, space and time consist of smallest finite-sized intervals, i.e., there exist a minimum time interval and a minimum space interval. They may be respectively called time unit and space unit. Modern physics implies that their values are approximately $1.1 \times 10^{-43} s$ and $3.2 \times 10^{-35} m$. These values are extremely small, so we never directly "see" the discreteness of space and time, and have been thinking space and time are continuous.

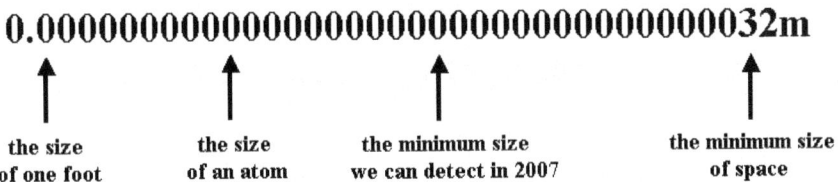

Figure 6.2 How far is it to the minimum size of space?

Can we detect the discreteness of space and time?

World's largest particle accelerator LHC (Large Hadron Collider) can approach a space about $10^{-20} m$ by colliding beams of protons at an energy of about 10 TeV, where 1TeV is about the energy of motion of a flying mosquito. But this tiny distance is still 10^{15} times larger than a space unit. Based on scaling, in order to approach the minimum space interval, we would need a particle accelerator the size of the galaxy or even the size of the current universe. Maybe we can never directly detect the minimum sizes of space and time. But logic can leads us to the deepest place of nature with the help of a little experience. As we will see, the convincing evidence of the maximum of the speed of light may have revealed the discreteness of space and time. Moreover, God plays observable dice with the universe in discrete space and time. Through detecting the process of God playing dice, we can also approach the minimum sizes of space and time.

Due to the discreteness of space and time, any physical being can but exist in a space region not smaller than the space unit, and any physical becoming can but happen during a time interval not shorter than the time unit.[5] As a result, a particle is no longer in one 0-sized position at a durationless instant (as in continuous space and time), but in a space unit during a time unit in discrete space and time. This defines the existent form of a particle in discrete space and time.

The analysis of motion in continuous space and time also applies to the motion in discrete space and time. In addition, the discreteness of space and time has more restrictions on the possible forms of motion. As we will see immediately, it actually requires the existence of discontinuous motion. Due to the limitation of discreteness of space and time, there are only two possible free motion states for continuous motion: one is rest state; the other is the motion state with a constant speed, which equals to space unit divided by time unit, and turns out to be just the speed of light $3 \times 10^8 m/s$.

If the speed of an object is larger than the speed of light, the object will move more than a space unit during a time unit. Then moving a space unit will correspond to a time interval shorter than a time unit during the continuous movement. This contradicts the discreteness of space and time, which requires that a time unit is the minimum time interval.[6]

On the other hand, if the speed of an object is smaller than the speed of light, the object will move a space unit during a time interval longer than a time unit. Then during a time unit the object will move a space interval shorter than a space unit during the continuous movement. This also contradicts the discreteness of space and time. So, a free object can only be still or move with the speed of light in discrete space and time if its motion is continuous. This result is obviously inconsistent with experience. A free object can move with a speed different from the speed of light in reality.[7]

Therefore, the motion of objects will be not continuous but discontinuous and random if space and time are discrete. In short, God must play dice in discrete space and time. This is a new surprise; the discreteness of space and time actually requires the existence of random discontinuous motion. But is the played dice observable? In the following, we will illustrate that the discreteness of space and time can indeed result in the emergence of the inherent randomness of motion through a random collapse process of the particle cloud, which is also called the dynamical collapse of the wave function. So, God does play observable dice in discrete space and time.

Consider a superposition of two still particle clouds (i.e. stationary states) with different energies. The two branches are mainly separated in space but also have a small overlap. According to the linear Schrödinger evolution, the density of the superposed particle cloud will oscillate in the overlapping region with a period inversely proportional to the energy difference of the two stationary states. When the energy difference is very small, this oscillation seems to cause no problem. However, when the energy difference is very large, the period of oscillation will be extremely short. This may contradict the discreteness of space and time. For instance, if the energy difference exceeds the so-called Planck energy, which is approximately the energy sum of 10^{19} hydrogen atom, the period of oscillation will be shorter than a time unit.[8] But this is impossible, as the time unit is the minimum time interval in discrete space and time, and no change can happen during a time interval shorter than it.

So, the superposed particle cloud with energy diffusion larger than Planck energy cannot exist, and must have collapsed into one of the branches in the superposition before the energy diffusion reaches such a large value because of the requirement of discrete space and time. Due to the randomness of motion, the collapse process of a particle cloud is also random, and its outcome just releases the inherent randomness of motion.

The above example clearly shows that the discreteness of space and time will result in the observable random collapse of a particle cloud. Moreover, the minimum sizes of discrete space and time also yield a plausible collapse criterion. It is that if the energy diffusion of a particle cloud approaches the Planck energy, the particle cloud will collapse into one of the definite energy branches in about a time unit.[9]

The collapse criterion also implies that when the energy diffusion of a particle cloud is smaller than the Planck energy, it will take a longer time for the particle cloud to collapse. This means that the random collapse process of a particle cloud is generally gradual. Then what are the details of the random collapse? How in hell does God play observable dice? Is God's gambling rule really fair-and-square?

The gravity-induced wavefunction collapse conjecture

The origin of the wavefunction collapse is still a deep puzzle. It might be natural to guess that the wavefunction collapse is induced by gravity. The reasons include: (1) gravity is the only universal force existing in all physical interactions; (2) gravitational effects grow with the size of the objects concerned, and linear superpositions is violated only in the context of macroscopic objects.

The gravity-induced collapse conjecture can be traced back to Feynman. In his *Lectures on Gravitation* published in 1960s, Feynman considered the philosophical problems in quantizing macroscopic objects and contemplates on a possible breakdown of quantum theory. He said, "I would like to suggest that it is possible that quantum mechanics fails at large distances and for large objects, ...it is not inconsistent with what we do know. If this failure of quantum mechanics is connected with gravity, we might speculatively expect this to happen for masses such that ... 10^{-5} grams." [10]

The British physicist Roger Penrose further strengthened the gravity-induced collapse argument in 1990s.[11] He argued that, due to the fundamental conflict between the general covariance principle of general relativity and the superposition principle of quantum mechanics, the superposition of different space-times is physically improper, and the evolution of such a superposition cannot be defined in a consistent way. This requires that a quantum superposition of two space-time geometries, which corresponds to two macroscopically different energy distributions, should collapse after a very short time. It is expected this interesting conjecture could be tested in the near future.

God's gambling rule

> *We know that his play is always fair, just, and patient.*
> —Thomas Huxley

Although some details of the random collapse of a particle cloud are still unknown, we can give a basic picture of this process and its general law. As we will see, God not only plays observable dice with the universe, but also has a fair gambling rule.

Consider again the energy-superposed particle cloud discussed in the last section. The particle stochastically stays in one energy branch during each time unit. To begin with, the probability of the particle staying in each branch is proportional to the density of that branch. It means that the particle will stay for a longer time in the branch with larger density. This is the first law of random collapse, namely the first part of God's gambling rule. It is uniquely fixed by the existence of random discontinuous motion.

Next, the random stay in one branch will change the density of the branch. Concretely speaking, it will increase the density of the branch. The density increase is proportional to the summation of the densities of other branches, and the coefficient is related to the energy diffusion of the particle cloud.[12] Correspondingly, the densities of other branches will be scaled down. This is the second law of random collapse, namely the second part of God's gambling rule. Its form is uniquely fixed by the requirement that the particle cloud as an actual existence can present itself exactly.

The random collapse process of a particle cloud will be completely determined by these two laws. As a result, the density of the particle cloud will undergo a stochastic collapse evolution during a time interval much longer than a time unit. The collapse outcome will be that the particle cloud stochastically collapses into a determinate energy branch, and the collapsing probability is proportional to the initial density of that branch. Besides, the collapse time is inversely proportional to the square of the energy diffusion of the particle cloud. In this way, the inherent randomness of motion is finally released through the observable random collapse outcome. Moreover, the particle cloud will also present itself through the distribution of the collapse outcomes in an ensemble consisting of a large number of identical particle clouds.

The above two laws constitute the whole gambling rule of God, according to which He plays observable dice with the universe. In the following, we will give a graphic description of God's gambling rule.

Suppose there are two gamblers Alice and Bob. Each time a special coin is tossed up to determine who wins. The probability of the coin to be each side is adjustable.[13] The gambling rule is as follows.[14] If the coin is heads, then Alice win, otherwise Bob wins. The probability of the coin to be heads is proportional to Alice's current stake, while the probability of the coin to be tails is proportional to Bob's current stake. The rule means that the winning probability of each gambler is proportional to his (or her) current stake each time. This is just the first part of God's gambling rule. Besides, the loser gives a fixed proportion of his (or her) current stake to the winner each time. This is the second part of God's gambling rule.

Let Alice's initial stake be 80 dollars, and Bob's initial stake be 20 dollars. The fixed paying proportion is 1/10. At the beginning, the winning probability of Alice is 80%, and the winning probability of Bob is 20%. If the coin is heads, then Alice wins, and Bob will give 1/10 of his current stake, namely 2 dollars, to Alice. Then Alice's current stake is 82 dollars, and Bob's current stake is 18 dollars.

They continue to gamble. Now the winning probability of Alice is 82%, while the winning probability of Bob is 18%. If this time the coin is tails, then Bob wins, and Alice will give 1/10 of her current stake 82 dollars, namely 8.2 dollars, to Bob. Then Alice's current stake is 73.8 dollars, and Bob's current stake is 26.2 dollars. The gamble can continue in this way until one gambler loses all his (or her) stakes.

Then what is the probability of each gambler wining all stakes? A simple calculation shows that this probability is proportional to the initial stake of each gambler. As thus, the probability of Alice wining all stakes is 80%, and the probability of Bob wining all stakes is 20%. This result indicates that God's gambling rule is fair-and-square. The more stakes, the more probability to win. But the gambler with fewer stakes still has chance to win. Further calculation shows that the average gambling times is inversely proportional to the square of the fixed paying proportion. Besides, it also depends on the minimum unit of the stakes and the rounding rule; the smaller the minimum unit, the more the gambling times.

As to God's actual gamble, the gamblers are not Alice and Bob, but two energy branches of a particle cloud. At each time unit, the gamble proceeds one time. The stake of each branch is its density. No special coin is needed. The winning or losing of each branch is naturally determined by the random stay of the particle in the branches. The branch where the particle stays will win, and the other branch will lose. The winning probability of each branch is proportional to its current density. The "paying" proportion of the losing branch is proportional to the energy diffusion of the particle cloud. This gamble is the fastest action in the universe; it proceeds one time during each time unit, the shortest time interval in the universe. Aha, we have reached the deepest level of reality here.

If you have time, you can gamble with your friend by God's gambling rule, which is absolutely fair. The game may have a strange fascination for you. Remember what you experience is just the incessant random dance of nature. Every thing in the universe, whether it is near or far, whether it is an atom or a star, dances in such a graceful way.

The unification of two worlds

> *It is a wonderful feeling to recognize the unity of a complex of phenomena that to direct observation appear to be quite separate things.*
> —Albert Einstein

As we have seen, random discontinuous motion requires the existence of discrete space and time to release its randomness; on the other hand, the discreteness of space and time not only leads to the existence of random discontinuous motion, but also can indeed release the inherent randomness of motion through a random collapse process of particle cloud. As a consequence, the actual motion will be random discontinuous motion in discrete space and time.

The picture of the actual motion of a particle is as follows. A particle stays in a space unit during a time unit. Then it will still stay there or stochastically appear in another space unit, which is probably very far from the original one, during the next time unit. During a time interval much larger than a time unit, the particle will move throughout the whole space to form a particle cloud with a certain density and flux density.

Although the complete law of motion is not available now, we can give its general form according to our previous analysis. The complete equation of motion will be a revised Schrödinger equation that contains two kinds of evolution terms of the wave function.[15] The first is the deterministic linear evolution term as in the Schrödinger equation, and the second is the stochastic nonlinear evolution term, which results in the random collapse process of the particle cloud (i.e. the dynamical collapse of the wave function). The equation is essentially a discrete one, and all quantities are defined in discrete space and time.

In the equation of motion, the deterministic linear evolution term will lead to the wave-like behaviors of the particle cloud such as interference, while the stochastic nonlinear term will lead to the particle-like behaviors of the particle cloud such as localization. Accordingly the evolution of a particle cloud will be a certain combination of the wave-like process and the particle-like process. Moreover, the relative strength of these two processes is determined by the energy diffusion of the particle cloud.

If the energy diffusion of the particle cloud is very tiny, then its collapse time will be extremely long, e.g. longer than the age of our universe. So the evolution of the particle cloud will be dominated by the linear Schrödinger evolution. This is what happens in the microscopic world, where the behaviors of microscopic particles are wave-like. For instance, in the double-slit experiment, a particle cloud passes through two slits at the same time, and the passed branches superpose and "interfere" with each other. Then after a large number of particle clouds pass through the two slits, they can form the double-slit interference pattern.

If the energy diffusion of the particle cloud is very large, its collapse time will be extremely short, e.g. shorter than the time interval during which even light can but travel one meter. Then the evolution of particle cloud will be dominated by the stochastic collapse evolution. As we will see, this is just what happens in the macroscopic world, where the behaviors of macroscopic objects are particle-like.

For a macroscopic object, its random discontinuous motion also forms an object cloud spreading throughout the whole space. For instance, a ball is also a ball cloud in reality. However, the environmental influences such as thermal energy fluctuations will result in the extreme largeness of the energy diffusion of the object cloud. Then the spreading of the object cloud will be greatly suppressed, and its evolution will be dominated by the collapse process or localizing process. This localizing process proceeds very frequently, and thus the density of the object cloud will always concentrate in a very small local region. So, a macroscopic object will be in a local region at each instant, and can but be approximately still or move continuously. This is just the appearance of continuous motion in the macroscopic world. As a result, we only see a living cat or a dead cat, and we never see Schrödinger's cat that is half alive and half dead. We never see a cat passing through two doors at the same time either.

Besides, the above localizing process will also result in the appearance of definite measurement outcomes of microscopic particles. For example, in the double-slit experiment, when a particle cloud passes through two slits and then reaches a macroscopic measuring apparatus (e.g. a detecting screen), the linear Schrödinger evolution of the whole system will generate a superposition of different outcome states of the system, in each of which the particle is measured in one definite position. But due to the extreme largeness of the energy diffusion of the superposition, which is introduced by the macroscopic measuring apparatus, the superposition will immediately collapse to one of the definite outcome states. As a result, the apparatus cloud as well as the particle cloud will localize in a definite position. So, a nonlocal particle cloud is detected only in a local position (e.g. a point on the detecting screen). This explains the other facet of the wave-particle dilemma, i.e., that a particle cloud also behaves like a local particle when it is measured.

Lastly, the law of the apparent continuous motion, namely Newton's laws of motion, can also be derived from the complete law of actual motion as an approximation. Besides, there should also exist a stochastic term in Newton's equation of motion, which stems from the stochastic nonlinear term in the complete equation of motion. Although this term is very small for most situations, it may be detected by precise experiments.

In a word, random discontinuous motion in discrete space and time provides a uniform realistic picture for the microscopic and macroscopic worlds. It is *the* real motion. Every thing in the universe, whether it is an atom or a ball or even a star, undergoes such motion every moment. The most familiar continuous motion is only its approximate display in the macroscopic world.

7
The Prime Mover

The deeper one penetrates into nature's secrets,
the greater becomes one's respect for God.
—Albert Einstein

Is it logical that God plays dice with the universe? Why does God play dice? Does this indicate that God has freedom or He must do so? In this chapter, we will go into the whys and wherefores. Our efforts are rewarding; we will finally understand God's thoughts here.

Causality and chance

The most incomprehensible thing about the universe
is that it is comprehensible.
—Albert Einstein

Every event has a cause. This is the common sense of causality. It is also called principle of causality, which has been one of the most influential beliefs in science and philosophy. As Immanuel Kant stated, causality is the basis of all scientific work, and it is the condition that renders science possible.[1] Especially, causality makes the world comprehensible. Many believed that reason could provide an absolute justification for this law, while others, notably Hume, argued that logic is incapable of providing a foundation in reason for the principle of causality. According to Hume, causes contain nothing within themselves that could enable them to act on anything else, and thus cause cannot logically necessitate effect.[2]

On the other hand, indeterminism, which is the doctrine that there are some events which have no cause, also has a long history. It can be traced back to Epicurus, according to which causality is limited by the spontaneous "swerve" of atoms that happens purely by chance. With the discoveries in the realm of quantum physics, especially Heisenberg's uncertainty principle, most people begin to accept that the principle of causality must be given up in the atomic world, and God indeed plays dice with the universe. However, the basis of indeterminism still needs to be established by logical considerations.

As Einstein remarked, the most incomprehensible thing about the universe is that it is comprehensible. The principle of causality is comprehensible, but it seems to be false in the actual world. On the other hand, indeterminism is probably real, but it appears to be irrational and incomprehensible. Looking for causes is a natural intelligent reaction. It seems ridiculous to answer that there is no cause or no reason resulting in the observed events — that events simply happen like that. Those who insist on the comprehensibility of nature, notably Einstein, uncompromisingly reject indeterminism, while the pragmatic hail indeterminism as one of the most profound discoveries of modern science, although they don't understand it either. This generates a huge schism in the scientific community. Then who is right? How to choose between the principle of causality and indeterminism for a rational man? The dilemma must be solved if the universe can be understood by reason.

According to the principle of causality, no events or changes can happen without cause, while indeterminism asserts that there are some events which have no cause. At first glance, these two beliefs seem utterly incompatible. As we will see, however, they can be unified into a generalized principle of causality. What's more, indeterminism is as logical and comprehensible as the principle of causality.

It is logical that a cause results in a lawful change, whereas a random change requires no cause. Then if there is no cause, there should exist two possible effects: one is that no change happens, the other is that a random change happens. The former accords with the existing principle of causality, while the latter is still possible in logic as we will see below. First, since a random change also requires no cause, the effect of random change cannot be logically excluded. Next, we can always assume the existence of a universal cause that results in the happening of random changes. Such a universal cause, which is irrelevant to time, cannot determine a concrete change, and thus the change must be random. As a consequence, the happening of random changes without cause may be logical from stem to stern.

Lastly, it is possible that random changes must happen in some situations due to a certain universal cause. Although the happening of random effects without cause is logical, it still needs to be determined that which of the above two possible effects, namely no change and a random change, will happen in an actual situation. If the random change must happen due to a certain universal cause, then it will be impossible that no change happens when there is no cause. This will determine which of the two possible effects will happen in reality.

In a word, even if no concrete cause exists, a change can still happen as long as the change is purely random. In order to further understand this conclusion, it is necessary to distinguish two kinds of causes. One is concrete causes that relate to time, and the other is universal causes that are irrelevant to time. The former is our familiar causes appearing in the principle of causality. Such a concrete cause will result in a lawful change at a concrete time. The latter is a new kind of causes, which are similar to Aristotle's final causes. A universal cause can result in ceaseless random changes. As a consequence, both lawful changes and random changes have their causes.

So, the principle of causality and indeterminism can be unified in a generalized principle of causality. According to the new law, there are two kinds of causes: concrete causes and universal causes, and accordingly there are two kinds of events: lawful events and random events. As an inference, a real event will generally have both a lawful element and a random element, so does the law of nature.

To sum up, we find an appealing solution to the long-standing puzzle of indeterminism. The existence of uncaused events is actually logical. So it is comprehensible that God plays dice with the universe. Moreover, the principle of causality and indeterminism can be unified in a generalized principle of causality. The new law will provide a rational basis for our everyday life as well as for science and philosophy. Maybe even Einstein would be happy about this outcome, as he was a rational man.

Why does God play dice?

Every why has its wherefore.
—An old saying

It is comprehensible that God plays dice with the universe; why change or motion is random is because it has no concrete cause. But if there is no cause at all, objects can either stand still or move in a random way. Then why do objects move in a random way when there is no concrete cause? Why in hell does God play dice? For everything which occurs there should be a reason or explanation for why it occurs, and why this way rather than that.[3] So why God plays dice should also have a universal cause or an ultimate reason. Let us find it now.

If an object can move only as a result of a concrete cause such as the external force exerted by other objects, the object would not be able to move without this cause, but, on the other hand, the cause cannot be generated if there is no motion, which is needed to transfer it between objects.[4] Thus either everything is immobile or there exists uncaused, random motion.

In a word, if objects didn't move randomly when there is no concrete cause (e.g. force or interaction), all objects would be resting, and all interactions between objects would also be non-existent. It seems that Zeno's motionless world come back again like a half penny.

Is such an absolutely still world possible? The answer may be negative. The existence of an object is represented by its properties such as mass and charge, and these properties are defined by its interaction with other objects. If there were no interactions, then all objects would be devoid of their properties, and would not exist either. Thus, an absolutely motionless world cannot exist in all probability.

So, it seems that objects must move spontaneously and randomly in order to exist. This means that the universal cause for the random motion of objects is probably the existent need of objects. As thus, the ultimate reason why God plays dice is that He must do so in order to make the world exist.

Who is God?

> *Practice not-doing, and everything will fall into place.*
> — Lao Tzu, *Tao Te Ching*

God plays dice with the universe. Moreover, He must do so in order to make the universe exist. Then who is God? This is the last question. It is widely accepted by scientists that God is an anthropopathic appellation of the universe. This is also its original meaning when Einstein said God does not play dice. Yet such a viewpoint seems too naïve. As we will see, random motion will reveal more truth about the Old One.

With the development of science, the role of God has been constantly changing. If the universe could be wholly understood by science, then God would seemingly have no position. Yet this end can never be reached. So God always has His position in human world.[5]

In Aristotle's physical world, God has his position. The medieval theologian St. Thomas Aquinas gave a proof of God's existence based on Aristotle's *Physics*. According to Aristotle, external force is the cause of motion; moving objects only continue to move when there is an external force inducing them to do so. So every moving object needs a mover. This fact, to Aquinas, shows that God, which is defined as the First Mover, exists. He argued, "this (i.e., the fact that one is moved by another) cannot go on to infinity, because then there would be no first mover, and consequently no mover, ... Therefore it is necessary to arrive at a first mover, moved by no other; and this everyone understands to be God."[6]

In Newton's physical world, God has a new position. According to Newton's first law of motion, an object can sustain its motion when no external force is applied to it. A moving object needs no mover. So there is no need for Aquinas's First Mover. However, Newton's First Mover still exists. According to the other half of the first law of motion, every object remains at rest until moved by another object. No object has the ability to move itself. Then who moved the first moving object? How did it start off if no object can move itself?[9] So, as Newton thought, the universe still needs some original thing that set it all in motion at the beginning. This new First Mover, for Newton, was God. Indeed, Newton warned against using his mechanics to view the universe as a mere machine, like a great clock. He said, "Gravity explains the motions of the planets, but it cannot explain who set the planets in motion. God governs all things and knows all that is or can be done."[8]

Today we know both Aristotle and Newton are wrong. So their Gods no longer exist. According to the new picture of random motion, objects can move by themselves. Hence it seems that the First Mover is no longer needed. What is the position of God in the new universe then?

Imagine the picture of random motion. An object is in one position at an instant. Then it randomly appears in another position at next instant.

No concrete cause results in the random change of its position. However, if there is no cause at all, objects can either stand still or move in a random way. So there must exist a universal cause inherent in all objects that results in their random motion. As we have seen in the last section, this universal cause may be the existent need of objects; objects must move in a random way in order to exist. So God seems to have no position in the spontaneous universe. If God did exist, He would need to do nothing. In the profound words of the great Chinese sage Lao Tzu, "Practice not-doing, and everything will fall into place."[9] This is the very Tao of the universe.

Notes

Chapter 1

1. See, for example, W. C. Salmon (eds.), *Zeno's Paradoxes* (New York: Bobbs-Merrill, 1970). For a recent popular introduction, see J. Mazur, *The Motion Paradox: The 2500-Year-Old Puzzle Behind the Mysteries of Time and Space* (New York: Dutton, 2007).
2. René Descartes is regarded as the father of modern philosophy partly due to this famous sentence. It first appeared in Descartes' philosophical classic *Discourse on Method*, published in 1637.
3. David Hume is probably the most important philosopher ever to write in English. He is most well-known for his penetrating analysis of causation. Based on this analysis, he famously argues that our belief that the sun will rise tomorrow has no rational basis. We will refer to Hume's analysis of causation in the last chapter of this book.
4. The aphorism is in the Analects of Confucius. See, for example, Confucius, *The Analects*, trans. D. C. Lau (London: Penguin, 1979).
5. I. Barrow, *Mathematical Lectures* (London, 1734), p.143.
6. See F. Arntzenius, "Are there really instantaneous velocities?", *The Monist* 83 (2000), 187-208.
7. H. Bergson, *Creative Evolution*, trans. A. Mitchell (New York: Henry Holt, 1911), p.308.
8. B. Russell, *Principles of Mathematics* (Cambridge: Cambridge University Press, 1903), p.473.

Chapter 2

1. Su Tungpo (1035-1101) was an eminent poet, painter, and statesman of the Northern Song Dynasty of China. See Y. T. Lin, *The Gay Genius: The Life and Times of Su Tungpo* (Greenwood Press, 1971).

2. R. Feynman, *The Character of Physical Law* (London: Penguin, 1992), p.129.

3. Quoted in W. Heisenberg, *Physics and Beyond* (New York: Harper and Row, 1971), p. 206.

4. The demonstration of the orthodox view appears to be flawless, and it indeed dusted the eyes of nearly all great men in the twentieth century. The underlying reason, as we think, is that people including Einstein and Bohr all held the following ingrained prejudice, i.e., that continuous motion is the only possible form of motion. To learn the deadly flaws in the demonstration, see S. Gao, *Quantum: A Historical and Logical Journey* (Beijing: Tsinghua University Press, 2003) and S. Gao, *Quantum Motion: Unveiling the Mysterious Quantum World* (Bury St Edmunds: Arima Publishing, 2006).

5. The debate between Einstein and Bohr was called the "battle for the soul of physics" by Andrew Whitaker. For a detailed discussion, see A. Whitaker, *Einstein, Bohr and the Quantum Dilemma* (Cambridge: Cambridge University Press, 1996).

6. For a full exposition of Bohm's view, see D. Bohm and B. J. Hiley, *The Undivided Universe: An Ontological Interpretation of Quantum Theory* (New York: Routledge, 1993).

Chapter 3

1. See, for example, J. Clement, "Students preconceptions in introductory mechanics", *American Journal of Physics*, 50 (1982), 66-71.

2. According to Thomas Kuhn, an American philosopher of science, the change was a Gestalt switch or paradigm transformation. See T. S. Kuhn, *The Structure of Scientific Revolutions* (Chicago: University of Chicago Press, 1996).

3. Quoted in R. S. Westfall, *Never at Rest: A Biography of Isaac Newton* (Cambridge: Cambridge University Press, 1983).

4. Ibid.

5. Ibid.

6. In classical mechanics, the definition of velocity is that the instantaneous velocity for an object is the limit of the object's average velocity as the time-interval around the instant in question tends to zero. As thus, velocity is not an instantaneous intrinsic property of objects in essence.

7. As we will see later, velocity is only an approximate and average display of motion,

and it does not exist in a strict sense for the motion of objects, which is random and discontinuous in nature.

8. For a strict proof of this conclusion see S. Gao, *Quantum Motion: Unveiling the Mysterious Quantum World* (Bury St Edmunds: Arima Publishing, 2006).

9. During his first visit to Princeton University in 1921, Einstein once said to the mathematician Oswald Veblen, "Subtle is the Lord, but malicious He is not." This remark is famously quoted in A. Pais, *Subtle Is the Lord: The Science and the Life of Albert Einstein* (Oxford: Oxford University Press, 1983).

Chapter 4

1. Einstein made this famous remark in a letter to Max Born dated 4 December 1926. Quoted in W. Isaacson, *Einstein: His Life and Universe* (New York: Simon & Schuster, 2007).

2. Einstein wrote this sentence in a letter to Borns dated 29 April 1924. See M. Born, *The Born–Einstein Letters 1916–1955: Friendship, Politics and Physics in Uncertain Times* (New York: Macmillan, 2005), p.79.

3. This consequence also implies that a free object cannot stay in the same position all the time, i.e., that a free object cannot be absolutely motionless.

4. See I. Newton, *Opticks*, I. B. Cohen *et al* (eds.), (New York: Dover, 1952).

5. J. Jeans, *Physics and Philosophy* (Cambridge: Cambridge University Press, 1943), p.149-150.

6. Quoted in M. Kline, *Mathematical Thought from Ancient to Modern Times* (Oxford: Oxford University Press, 1990), p. 973.

7. This mathematical revolution transformed classical analysis into modern analysis. It is very similar to the physical revolution happening in the same period, which transformed classical mechanics into quantum mechanics. As we will see, there indeed exists an important connection between them. If this connection were noticed then, the course of science in the twentieth century would be greatly different.

8. Quoted in M. Kline, *Mathematical Thought from Ancient to Modern Times* (Oxford: Oxford University Press, 1990), p. 973.

9. See, for example, W. G. Englert, *Epicurus on the Swerve and Voluntary Action* (Atlanta: Scholars Press, 1987).

10. For a detailed discussion of Al-Nazzam's leap motion, see, for example, M.

Jammer, *The Philosophy of Quantum Mechanics: The Interpretations of Quantum Mechanics in Historical Perspective* (New York: John Wiley & Sons, 1974).

11. Quoted in M. Jammer, *The Philosophy of Quantum Mechanics: The Interpretations of Quantum Mechanics in Historical Perspective* (New York: John Wiley & Sons, 1974).

12. Before we further understand random discontinuous motion, it may be helpful to first answer the following question. If motion is indeed discontinuous and random, then why has nobody discovered this fact? Even though the idea of random discontinuous motion is purely nonsense, why has nobody even referred to or refuted it? In order to answer this question, we have to search for the most deep-rooted prejudices in human mind, which have been preventing people from understanding the real nature of space-time and motion. As Galileo claimed, not knowing the motion, nature too is unknown. Indeed, the law of motion has been being deeply studied with the help of experience. But unfortunately, the study of motion itself seems to be badly ignored in physics. Especially, as we have seen, Newton's revolution of classical mechanics unexpected terminated the study. The orthodox slogan is "we need only explain changes in motion, not motion itself." This neglect finally resulted in a serious aftermath in the twentieth century, namely the unprecedented chaos in the understanding of quantum theory. As we think, why nobody understands quantum mechanics is mainly because nobody understands motion itself. In the final analysis, the quantum dilemma lies in that continuous motion cannot explain quantum phenomena, but it is taken for granted that continuous motion is the only possible form of motion. So, Einstein could not accept quantum because it excludes the only reality he cherished, continuous motion, while Bohr would rather reject reality because the only reality, continuous motion, is excluded by the quantum he trusted. In a similar way, why Feynman took the double-slit experiment as the only mystery is because continuous motion cannot explain it, but continuous motion was unconsciously regarded by him as the only possible motion. As thus, it is the ingrained prejudice, i.e. that continuous motion is the only possible form of motion, that has been preventing people from understanding the meaning of quantum theory, as well as the real nature of space-time and motion. Today nearly all people agree that continuous motion can not explain the double-slit experiment. Notably, the experiment was regarded by Feynman as "a phenomenon which is impossible, absolutely impossible, to explain in any classical way". However, although the picture

of continuous motion is utterly rejected, no one has seriously considered another form of motion different from continuous motion. In fact, the faint picture of such new motion had occasionally appeared, for example, in Bohr's model of atoms in the form of discontinuous quantum jump, as well as in Heisenberg's uncertainty paper in the form of discontinuous trajectory of a particle (Bohr 1985). Even Schrödinger, the strong opponent of quantum jump, had also tried to revise the classical concepts, especially space-time and causality, in order to reach a new realistic picture of motion (Moore 1989). But why has nobody suggested that motion is not continuous? It is just the above prejudice that prevents people to do so. Bohr's situation is a typical instance. At first glance, Bohr seemed to be more radical; he boldly took the inability of continuous motion as a reason to reject reality. Yet he also unconsciously held, even more strongly than Einstein, the same prejudice, namely that continuous motion is the only possible form of realistic motion. Therefore, he would rather play the shaky game with reality. Now, once the prejudice is rejected, everything immediately becomes so obvious; if motion cannot be continuous, it must be discontinuous. This is the simplest logic. As we will see immediately, after we have understood motion, we can then understand quantum mechanics. Motion is discontinuous and random in reality. Continuous motion is merely the shadow of the real motion. As thus, a single particle can pass through two slits at the same time in the double-slit experiment; it needs not to be divided, but only needs to move discontinuously. Aha, nature is logical and comprehensible; only the way we understand it may be unreasonable and impenetrable.

Chapter 5

1. For details see S. Gao, *Quantum Motion: Unveiling the Mysterious Quantum World* (Bury St Edmunds: Arima Publishing, 2006).
2. The word "particle cloud" has been used in different meanings by physicists. For instance, a particle cloud usually denotes an ensemble of particles. See the latter discussion about electron cloud.
3. The density of the particle cloud corresponds to the position measure density of the point set in math.
4. Since a basic cloud evenly spreads out in the infinite space, it doesn't exist in a practical situation.

5. The flux density of the particle cloud corresponds to the position measure flux density of the point set in math.

6. In the original, historic "double slit" experiment done by Young, these photon clouds were actually split with what Young described as "a slip of card, about one thirtieth of an inch in breadth (thickness)", and passed on its both sides at the same time.

7. In fact, this is only one facet of the wave-particle dilemma. In the next chapter, we will discuss the other facet, i.e., that the particle cloud also behaves like a local particle through a dynamical collapse process.

8. See S. Gao, *Quantum Motion: Unveiling the Mysterious Quantum World* (Bury St Edmunds: Arima Publishing, 2006).

9. The existence of two kinds of bases is essentially due to the discontinuity of motion. For continuous motion there is only one kind of local states. Velocity and position are both local properties, in which velocity equals to the first derivative of position with respect to time.

10. The requirement of information conservation discussed in the last section is just a consequence of this one-to-one relation.

11. It is worth noting that there doesn't exist a physically meaningful solution for the situation where the reverse transformation is the same as the transformation.

12. The original paired quantities in the transformation should be position and velocity (i.e. the velocity of the nonlocal basic cloud or momentum basis). Since the transformation is generally different for different particles, a coefficient must be introduced in the transformation to denote the given particle. This coefficient is now called mass of particle, and the product of mass and velocity is then defined as momentum.

13. For a recent popular introduction of Heisenberg's uncertainty principle, see D. Lindley, *Uncertainty: Einstein, Heisenberg, Bohr, and the Struggle for the Soul of Science* (New York: Doubleday, 2007).

14. Concretely speaking, the group velocity of the wave function describing the momentum basis of a particle equals to the momentum of the particle divided by the mass of the particle.

15. In Newton's classical mechanics, this relation is guessed with the help of experience. Here we deduce it by logic.

16. This derivation is not fundamental. In principle, we should derive the equation

from the basic form of interaction.

17. For a detailed introduction to quantum mechanics, see, for example, R. Shankar, *Principles of Quantum Mechanics* (Springer, 1994) and D. J. Griffiths, *Introduction to Quantum Mechanics (2nd Edition)* (Benjamin Cummings, 2004).

18. The figure is drawn according to L. I. Schiff's textbook *Quantum Mechanics* (New York: McGraw-Hill, 1968).

19. Quoted in W. Moore, *Schrödinger: Life and Thought* (Cambridge: Cambridge University Press, 1989).

20. The wave function is the most fundamental concept of quantum mechanics. A grasp of its meaning is crucial to achieving a genuine understanding of quantum mechanics from any perspective. At present, the orthodox interpretation of the wave function, which was first proposed by the German physicist Max Born, is that it is a probability wave giving the probability of finding a particle in a particular position, concretely speaking, the square of its amplitude represents the probability density for a particle to be measured in certain locations. On the other side, the wave function is taken as an objective physical field in some alternative realistic interpretations such as Bohm's theory. Yet, all existing interpretations of the wave function meet serious difficulties when they are further examined. Despite the formidable achievements of quantum mechanics, admittedly, there is no consensus what it tells us about Nature. Here we interpret the wave function as a mathematical description of the actual motion of particles, which is random and discontinuous in nature. The random discontinuous motion of particles is then taken as the quantum reality underneath the wave function. The conclusion may be inevitable when the probability relating to the wave function is not only the display of the measurement results, but also the objective character of the motion of particles, which means that the state described by the wave function is an objective indefinite state. Let us see what an indefinite state looks like if it has a picture in space and time. We take the indefinite position state of a particle as an illustration. A particle can but be in a definite position at each instant, as it has no time to move. Thus an indefinite position state cannot exist at instants, but exist in a time interval. Since an infinitesimal time interval near a given instant contains infinitely many instants, all possible positions in an indefinite position state can be distributed there. Moreover, the distribution of the positions at these instants can also be consistent with that of the random measurement outcomes of position at the given instant if these positions are random. In such an indefinite position state, the

particle is in one position at an instant, but at the instant immediately neighboring it is randomly in another position, which is probably very far from the previous one. As thus, the position change of the particle must be discontinuous and random. So, there is actually a logical road from the wave function to quantum reality. More than eighty years ago, Schrödinger wrote, "It has even been doubted whether what goes on in an atom can be described within a scheme of space and time. From a philosophical standpoint, I should consider a conclusive decision in this sense as equivalent to a complete surrender. For we cannot really avoid our thinking in terms of space and time, and what we cannot comprehend within it, we cannot comprehend at all."(Moore 1989) Now the picture of random discontinuous motion of particles in space and time might satisfy Schrödinger, the discoverer of the wave function and its equation. For a detailed discussion about the interpretation of quantum mechanics in terms of random discontinuous motion, see S. Gao, *Quantum Motion: Unveiling the Mysterious Quantum World* (Bury St Edmunds: Arima Publishing, 2006).

21. Schrödinger coined the term 'entanglement' to describe the peculiar connection between quantum systems in E. Schrödinger, "Discussion of Probability Relations Between Separated Systems," *Proceedings of the Cambridge Philosophical Society* 31 (1935), 555-563. He said: "When two systems, of which we know the states by their respective representatives, enter into temporary physical interaction due to known forces between them, and when after a time of mutual influence the systems separate again, then they can no longer be described in the same way as before, viz. by endowing each of them with a representative of its own. I would not call that *one* but rather *the* characteristic trait of quantum mechanics, the one that enforces its entire departure from classical lines of thought. By the interaction the two representatives [the quantum states] have become entangled."

22. For a recent popular introduction of quantum entanglement, see B. Clegg, *The God Effect: Quantum Entanglement, Science's Strangest Phenomenon* (New York: St. Martin's Press, 2006).

23. It is worth noting that the existence of a whole in the form of time division is not absolute for an entangled system. Its wholeness is relative to the entangled property of particles. The particles in an entangled state may be different kinds and thus can still be identified. It is still valid to talk about the respective existence of the particles in an entangled state.

24. As we will see in the next chapter, it is not because the cat died off, but because it

is half alive and half dead.

Chapter 6

1. J. S. Bell, "Against 'measurement'", *Physics World* 3 (1990), 33-40. In J. S. Bell, *Speakable and Unspeakable in Quantum Mechanics* (Cambridge: Cambridge University Press, 2004).

2. For instance, in the double-slit experiment, a particle cloud passing through two slits collapses into a local region of the detecting screen in a random and discontinuous way. We can only detect a local particle, not a nonlocal particle cloud.

3. The Greek epic poet Homer wrote this sentence in his epic *Iliad*, one of the greatest works in western literature. It may be the first linguistic bond between time and continuity. Today people still think time is continuous.

4. There are also some clues of the discreteness of space and time in modern physics. For instance, the appearance of infinity in quantum field theory and singularity in general relativity has implied that space and time may be not continuous but discrete. Besides, it has been widely argued that the proper combination of quantum theory and general relativity, two results of which are the formula of black hole entropy and the generalized uncertainty principle, may inevitably result in the discreteness of space and time.

5. The discreteness of space and time is essentially one kind of fuzzy property due to the universal existence of random discontinuous motion, and thus the space-times with a difference smaller than the minimum size are not absolutely identical, but nearly identical in physics.

6. It is worth noting that this argument also proves that the speed of light is the maximum speed of continuous process. This will explain the most mysterious and bewildering aspect of Einstein's special relativity, i.e., the maximum of the speed of light. As thus, the discreteness of space and time may provide a deeper logical foundation for special relativity. On the other hand, the maximum of the speed of light may have revealed the discreteness of space and time. For details see S. Gao, *Quantum Motion: Unveiling the Mysterious Quantum World* (Bury St Edmunds: Arima Publishing, 2006).

7. It can be conceived that the free object moves with the speed of light during some time units, and stays still during other time units. Then its average speed can be

different from the speed of light, and thus such motion can be consistent with the existing experience. However, the speed change of the free object during such motion can hardly be explained. In addition, such motion will contain some kind of unnatural randomness (e.g. during each time unit the speed of the free object will assume the speed of light or zero in a random way), which has no logical basis.

8. This is the total energy of our universe when it is about $10^{-43} s$ (i.e. a time unit) old. Now it is only about the total energy of a mosquito.

9. Since energy distribution determines space-time structure according to Einstein's general relativity, the criterion can be generally expressed by a comparison of two space-time sizes. It is that if the difference of the space-times in a superposition equals to a space unit, the superposition state will collapse to one of the definite space-times within a time unit The difference of space-times can be defined as the difference of the proper spatial sizes of the regions occupied by the branches in a superposition, which represents the fuzziness of the point-by-point identification of the spatial section of the space-times. For details see S. Gao, "A model of wavefunction collapse in discrete space-time", *International Journal of Theoretical Physics* 45(10) (2006), 1943-1957. Also in S. Gao, *Quantum Motion: Unveiling the Mysterious Quantum World* (Bury St Edmunds: Arima Publishing, 2006).

10. R. Feynman, *Feynman Lectures on Gravitation*, B. Hatfield (eds.), (Massachusetts: Addison-Wesley, 1995).

11. R. Penrose, "On Gravity's Role in Quantum State Reduction", *General Relativity and Gravity* 28 (1996), 581-600. For Penrose's recent idea about the wavefunction collapse, see R. Penrose, *The Road to Reality: A Complete Guide to the Laws of the Universe* (New York: Knopf, 2005).

12. The concrete formulation of this coefficient has not been found yet. For details see S. Gao, *Quantum Motion: Unveiling the Mysterious Quantum World* (Bury St Edmunds: Arima Publishing, 2006).

13. It may be difficult to make this special coin. Here we give a simple method of replacing the coin tossing. Make (or find) an even disk with a slick centre and mark degrees on its circumference. Put a rotatable scoop at its center. If it needs to generate an event with probability X% or 1-X%, mark a 360*X% degree sector on the disk, and then rotate the scoop by pushing its handle. If the handle stops inside the sector, then it indicates that an event with probability X% happened, otherwise an event with probability 1-X% happened. For example, mark a 360*80% (i.e. 288) degree sector

on the disk. Then if the handle stops inside the sector after it is rotated, it indicates that an event with probability 80% happened.

14. This game can be taken as a variation of gambler's ruin game, which was first used to describe the dynamical collapse of the wave function by the American physicist Philip Pearle.

15. This result implies that the existing quantum mechanics is not a complete theory. For a detailed discussion see S. Gao, *Quantum Motion: Unveiling the Mysterious Quantum World* (Bury St Edmunds: Arima Publishing, 2006).

Chapter 7

1. Kant held that causation is an *a priori* concept of the understanding. With regard to physics, he took as *a priori* the principle of causality besides space and time. See, for example, I. Kant, *Critique of Pure Reason*, trans. N. K. Smith (London: Macmillan, 1973).

2. Hume rejects the rationality of the law of causality, and removes the essential necessary connection from causality. In this way, he degrades causality to mere constant conjunction of events or regularity. This viewpoint was once strongly supported by Russell. He characterized the "law of causality" as a harmful "relic of a bygone age", and even urged the "complete extrusion" of the word 'cause' from the philosophical vocabulary. However, Hume's argument is not wholly convincible. If there were no causal connection, or, if one event does not logically necessitate another, then why does regularity of events exist? Probably there would not be any regularity. In response to this objection, Hume holds that natural laws are not certain but contingent. So there is indeed no reason why the regularity exists. This is a disastrous conclusion for rationalists. Especially, it makes the universe unintelligible. In fact, we can present a heuristic refutation of the viewpoint that natural laws are all contingent. Consider the proposition "All natural laws are contingent", which is also a law of nature. If the proposition is right, then all natural laws will be contingent. Since the proposition itself is a natural law, it is also contingent, not certain. This immediately leads to a contradiction. Thus the proposition "All natural laws are contingent" must be wrong. As a result, some natural laws must be certain, not contingent. This means that these laws have logical necessities, which can explain why the regularities in these laws exist. As we think, Hume's scope is limited by his

sheer empiricism; he removes necessary connexion from causality, but still keeps regularity in causality. If Hume were able to go beyond empiricism, he could reach a more satisfying conclusion. Now that one event cannot logically necessitate another, their connection cannot be causal and regular, but must be completely random. As thus, Hume's doubt about the law of causality can actually lead us to indeterminism, and his analysis can also make indeterminism or the existence of uncaused events logical and comprehensible. As we will see immediately, this will further help to provide a promising way to unify the law of causality and indeterminism.

3. The postulate is usually called principle of sufficient reason.

4. According to the existing theory of interaction, namely quantum field theory, the force or interaction between two particles is transferred by other particles. For instance, the electromagnetic interaction between two electrons is transferred by the so-called virtual photons.

5. For a recent discussion of God, see R. Dawkins, *The God Delusion* (New York: Houghton Mifflin, 2006) and V. J. Stenger, *God: The Failed Hypothesis. How Science Shows That God Does Not Exist* (New York: Prometheus, 2007).

6. T. Aquinas, *St Thomas Aquinas Summa Theologica*, trans. Fathers of the English Dominican Province (Christian Classics, 1981).

7. Aristotle had used this problem to refute Democritus's atomism, according to which the universe consists of atoms that generate all phenomenon by colliding into and combining with each other. Interestingly, he also solved the problem by positing an Unmoved Mover, which he called God. The Unmoved Mover was able to set things in motion without having to be moved itself at the beginning of time.

8. Quoted in J. H. Tiner, *Isaac Newton: Inventor, Scientist and Teacher* (Milford, Michigan: Mott Media, 1975).

9. This famous aphorism is in Tao Te Ching. See, for example, Lao Tzu, *Tao Te Ching, 25th-Anniversary Edition,* trans. G. F. Feng and J. English (New York: Vintage, 1997).

Bibliography

Aquinas, T. *St Thomas Aquinas Summa Theologica*, trans. Fathers of the English Dominican Province. Christian Classics, 1981.

Arntzenius, F. "Are there really instantaneous velocities?", *The Monist* 83 (2000), 187-208.

Barrow, I. *Mathematical Lectures.* London, 1734.

Bell, J. S. "Against 'measurement'", *Physics World* 3 (1990), 33-40.

Bell, J. S. *Speakable and Unspeakable in Quantum Mechanics.* Cambridge: Cambridge University Press, 2004.

Bergson, H. *Creative Evolution*, trans. A. Mitchell. New York: Henry Holt, 1911.

Bohm, D. and Hiley, B. J. *The Undivided Universe: An Ontological Interpretation of Quantum Theory.* New York: Routledge, 1993.

Born, M. *The Born–Einstein Letters 1916–1955: Friendship, Politics and Physics in Uncertain Times*. New York: Macmillan, 2005.

Bohr, N. *Collected Works* Volume 6, J. Kalckar (ed.) Amsterdam: North-Holland, 1985.Clegg, B. *The God Effect: Quantum Entanglement, Science's Strangest Phenomenon* New York: St. Martin's Press, 2006.

Clement, J. "Students preconceptions in introductory mechanics", *American Journal of Physics*, 50 (1982), 66-71.

Confucius, *The Analects*, trans. D. C. Lau. London: Penguin, 1979.

Dawkins, R. *The God Delusion*. New York: Houghton Mifflin, 2006.

Englert, W. G. *Epicurus on the Swerve and Voluntary Action.* Atlanta: Scholars Press, 1987.

Feynman, R. *Feynman Lectures on Gravitation*, B. Hatfield (ed.). Massachusetts: Addison-Wesley, 1995.

Feynman, R. *The Character of Physical Law*. London: Penguin, 1992.

Gao, S. "A model of wavefunction collapse in discrete space-time", *International Journal of Theoretical Physics* 45(10) (2006), 1943-1957.

Gao, S. *Quantum Motion: Unveiling the Mysterious Quantum World.* Bury St Edmunds: Arima Publishing, 2006.

Griffiths, D. J. *Introduction to Quantum Mechanics (2nd Edition)*. Benjamin Cummings, 2004.

Heisenberg, W. *Physics and Beyond*. New York: Harper and Row, 1971.

Isaacson, W. *Einstein: His Life and Universe*. New York: Simon & Schuster, 2007.

Jammer, M. *The Philosophy of Quantum Mechanics: The Interpretations of Quantum Mechanics in Historical Perspective.* New York: John Wiley & Sons, 1974.

Jeans, J. *Physics and Philosophy*. Cambridge: Cambridge University Press, 1943.

Kant, I. *Critique of Pure Reason*, trans. N. K. Smith. London: Macmillan, 1973.

Kline, M. *Mathematical Thought from Ancient to Modern Times*. Oxford: Oxford University Press, 1990.

Kuhn, T. S. *The Structure of Scientific Revolutions.* Chicago: University of Chicago Press, 1996.

Lao Tsu, *Tao Te Ching, 25th-Anniversary Edition,* trans. G. F. Feng and J. English. New York: Vintage, 1997.

Lin, Y. T. *The Gay Genius: The Life and Times of Su Tungpo.* Greenwood Press, 1971.

Lindley, D. *Uncertainty: Einstein, Heisenberg, Bohr, and the Struggle for the Soul of Science*. New York: Doubleday, 2007.

Mazur, J. *The Motion Paradox: The 2500-Year-Old Puzzle Behind the Mysteries of Time and Space.* New York: Dutton, 2007.

Moore, W. *Schrödinger: Life and Thought.* Cambridge: Cambridge University Press, 1989.

Newton, I. *Opticks*, I. B. Cohen *et al* (ed.). New York: Dover, 1952.

Pais, A. *Subtle Is the Lord: The Science and the Life of Albert Einstein.* Oxford: Oxford University Press, 1983.

Penrose, R. "On Gravity's Role in Quantum State Reduction", *General Relativity and Gravity* 28 (1996), 581-600.

Penrose, R. *The Road to Reality: A Complete Guide to the Laws of the Universe*. New York: Knopf, 2005.

Russell, B. *Principles of Mathematics.* Cambridge: Cambridge University Press, 1903.

Salmon W. C., ed. *Zeno's Paradoxes*. New York: Bobbs-Merrill, 1970.

Schiff, L. I. *Quantum Mechanics*. New York: McGraw-Hill, 1968.

Schrödinger, E. "Die gegenwärtige Situation in der Quantenmechanik", *Naturwissenschaften* 23 (1935), 807.

Shankar, R. *Principles of Quantum Mechanics.* Springer, 1994.

Stenger, V. J. *God: The Failed Hypothesis. How Science Shows That God Does Not Exist.* New York: Prometheus, 2007.

Tiner, J. H. *Isaac Newton: Inventor, Scientist and Teacher.* Milford, Michigan US: Mott Media, 1975.

Westfall, R. S. *Never at Rest: A Biography of Isaac Newton.* Cambridge: Cambridge University Press, 1983.

Whitaker, A. *Einstein, Bohr and the Quantum Dilemma.* Cambridge: Cambridge University Press, 1996.

Index

Al-Nazzam, Abu Ishaq Ibrahim, 46, 47, 106
Aquinas, Thomas, 101, 102, 115, 116
Aristotle, 7, 23, 24, 45, 98, 101, 102, 115
Arntzenius, Frank, 104, 116
Barrow, Isaac, 7, 104, 116
Basic cloud, vii, 52, 56, 57, 58, 59, 60, 108, 109
Becquerel, Henri, 39
Bell, John, 76, 112, 116
Bergson, Henri, 9, 104, 116
Blake, William, 35
Bohm, David, 20, 105, 110, 116
Bohr, Niels, 19, 105, 107, 109, 116, 117, 118
Borel, Emile, 43
Born, Max, 106, 110, 116
Buridan, Jean, 24
Causality, 23, 25, 29, 40, 95, 96, 97, 98, 99, 108, 114
    principle of, 95, 96, 97, 98, 99, 114
Cause
    of motion, 23, 24, 27, 28, 101
    universal, 29, 97, 98, 99, 100, 103
Chance, 80, 89, 95, 96

Classical mechanics, 26, 36, 40, 105, 106, 107, 109
Clement J., 105, 116
Collapse
    of the wave function, 83, 91, 114
Common sense, 5, 23, 25, 29, 74, 95
Confucius, xiii, 5, 104, 116
Continuity, 9, 14, 43, 77, 79, 112
Continuous
    motion, xi, 12, 13, 14, 15, 16, 17, 20, 33, 34, 41, 42, 44, 46, 52, 54, 59, 77, 82, 92, 93, 94, 105, 107, 109
    space and time, 77, 78, 80, 81, 82
Curie, Marie, 39
Democritus, 45, 115
Density, 50, 51, 52, 53, 54, 55, 57, 58, 60, 61, 65, 66, 72, 77, 83, 86, 87, 89, 90, 92, 108, 109, 110
    flux, 50, 53, 58, 61, 65, 66, 72, 77, 90, 109
Descartes, René, 4, 104
Discontinuity, 47, 50, 59, 69, 74, 75, 76, 77, 109
Discontinuous
    motion, vii, 34, 41, 42, 46,

47, 52, 59, 62, 68, 82, 90, 107, 111
Discrete
 discreteness, 76, 80, 81, 82, 83, 84, 90, 112
 space and time, vii, 76, 79, 80, 81, 82, 83, 84, 90, 91, 94
Double-slit experiment, 15, 16, 17, 18, 19, 20, 21, 22, 35, 54, 66, 91, 93, 107, 112
Durationless, 1, 5, 6, 7, 77, 81
Edison, Thomas, 48
Einstein, Albert, 1, 4, ix, 11, 20, 29, 31, 38, 56, 67, 90, 95, 96, 99, 101, 105, 106, 107, 109, 112, 113, 116, 117, 118
Electron cloud, 55, 70, 108
Entangled states, 72, 73
Entanglement, 73, 111
Epicurus, 45, 47, 96, 106, 116
Feynman, Richard, 15, 19, 49, 105, 107, 113, 116
Force, 3, 23, 24, 25, 26, 27, 28, 33, 52, 54, 69, 100, 101, 102, 115
Fourier transformation, 63, 64
Fourier, Joseph, 63, 64
Galilei, Galileo, 45, 107
Gambling rule, 77, 84, 86, 87, 89
Gao, Shan, 105, 106, 108, 109, 111, 112, 113, 114
Gunk, 7, 8

Half life, 40
Halley, Edmond, 26
Heisenberg, Werner, 63, 96, 105, 108, 109, 117
Homer, 77, 112
Hume, David, 4, 95, 104, 114
Huxley, Thomas, 86
Illusion, ix, 1, 2, 3, 4, 9, 13, 34, 35, 77
Impetus, 24, 25, 26
Indeterminism, 96, 97, 98, 99, 115
Inertia, 26, 27, 28
Infinitesimal, 8, 15, 42, 50, 51, 60, 77, 110
Interaction, 52, 66, 74, 100, 110, 111, 115
Interference, 59, 66, 91, 92
Jammer, Max, 107, 117
Jeans, James, 40, 106, 117
Jumping, xi, 12, 62, 76
Kant, Immanuel, 95, 114, 117
Kuhn, Thomas, 105, 117
Lao Tzu, 31, 59, 101, 103, 115
Laplace, Pierre Simon, 40
Leap, 46, 47, 106
Lebesgue, Henri, 43, 50
Leibniz, Gottfried Wilhelm, 11
Locality, 60, 62
Localizing process, 92, 93
Macroscopic
 objects, 15, 20, 34, 46, 76, 92

world, 35, 75, 77, 92, 94
Many minds, 21
Many worlds, 21
Measure, 43, 44, 50, 77, 108, 109
Microscopic
   particles, 15, 18, 20, 21, 35, 36, 41, 76, 78, 91, 93
   world, 35, 39, 41, 91
Momentum, 52, 60, 61, 62, 63, 64, 65, 109
Moore, Walter, 108, 110, 111, 117
Movie, 9, 10, 13
Mystery, 36, 39, 107
Newton, Isaac, ix, 7, 11, 20, 24, 26, 27, 28, 36, 37, 38, 52, 53, 54, 66, 93, 102, 105, 106, 107, 109, 115, 117, 118
Nonlocality, 60, 62
Occam's razor, 78
Ontology
   minimum, 7, 78
Pais, Abraham, 106, 117
Partial reflection, 35, 36, 37, 38, 53
Particle cloud, vii, 48, 49, 51, 52, 53, 54, 56, 57, 58, 59, 60, 61, 64, 65, 66, 67, 68, 69, 70, 72, 76, 79, 83, 84, 86, 87, 89, 90, 91, 92, 93, 108, 109, 112
   superposed, 57, 58, 59, 66, 83, 84, 86

Pascal, Blaise, 45
Pearle, Philip, viii, 114
Penrose, Roger, viii, 113, 117
Photon, 39, 53, 54, 73, 109
Planck, Max, 63, 64, 65, 83, 84
Plato, ix, xiv, 13
Point set, 6, 42, 44, 45, 50, 51, 108, 109
Prime Mover, 95
Probability, 5, 44, 55, 86, 87, 88, 89, 100, 110, 113
Psi wave, 68
Quantum, xiii, 48, 66, 73, 96, 106, 107, 110, 111, 112, 114, 115
   mechanics, 66, 106, 107, 110, 111, 114
Radioactivity, 39
Random
   changes, 97, 98
   collapse, 83, 84, 86, 87, 90, 91
   discontinuous motion, xii, 33, 34, 35, 41, 45, 47, 51, 62, 64, 66, 68, 74, 77, 79, 83, 86, 90, 92, 94, 107, 110, 112
   randomness, 35, 39, 41, 45, 47, 73, 76, 77, 78, 79, 80, 83, 84, 87, 90, 113
Relativity
   general, 112, 113
   special, 112
Russell, Bertrand, 10, 104, 114,

117
Rutherford, Ernest, 39, 40
Schrödinger equation, 66, 67, 68, 76, 77, 90
Schrödinger, Erwin, vii, 49, 56, 66, 67, 68, 72, 73, 75, 76, 77, 83, 90, 91, 92, 93, 108, 110, 111, 117
Shakespeare, William, 5
Space unit, 80, 81, 82, 90, 113
Speed of light, 82, 112
Spontaneous motion, 29
Stationary state, 54, 69, 83
Su, Tungpo, 11, 104, 117
Superposition, 57, 59, 60, 66, 83, 84, 93, 113
Swerve
    atomic, 45, 47, 96
Synchronism, 73, 74
Tao, xiii, 31, 101, 103, 115, 117
Tao Te Ching, 31, 101, 115, 117
Temporal atoms, 8
Time division, 48, 68, 70, 71, 74, 75, 111

Time flow, 70
Time unit, 80, 81, 82, 84, 86, 87, 89, 90, 112, 113
Trajectory, vii, 6, 11, 14, 21, 33, 52, 108
Transition, 10, 34
Uncertainty principle, 64, 96, 109, 112
Velocity, 12, 25, 27, 28, 32, 46, 52, 53, 60, 65, 105, 109
Wave function, 64, 65, 66, 67, 72, 76, 77, 91, 109, 110
Wavefunction collapse, 113, 116
Wave-particle dilemma, 56, 93, 109
Westfall, R. S., 105, 118
Whole
    inseparable, 48, 68, 70, 73, 74, 75
Wholeness, 48, 74, 75, 111
Yin-Yang, vii, 63
Young, Thomas, 54, 109
Zeno, 1, 2, 3, 5, 7, 8, 10, 27, 46, 100, 104, 117

OTHER WORKS BY SHAN GAO

Quantum Motion and Superluminal Communication (in Chinese)

Quantum: A Historical and Logical Journey (in Chinese)

Quantum Motion: Unveiling the Mysterious Quantum World

www.ingramcontent.com/pod-product-compliance
Ingram Content Group UK Ltd.
Pitfield, Milton Keynes, MK11 3LW, UK
UKHW021321180426
11947UKWH00015D/1056